FROM JARS TO THE STARS

How *Ball* Came to Build a Comet-Hunting Machine

Todd Neff

*Earthview*media

2010

Published by
Earthview Media
Denver, Colorado
info@earthviewmedia.com

Visit www.EARTHVIEWMEDIA.com.

From Jars to the Stars: How Ball Came to Build a Comet-Hunting Machine
© Todd Neff
All rights reserved
Printed and bound in the U.S.
First edition
ISBN-13: 978-0-9829583-0-8

This is a work of nonfiction. Any errors of fact will be corrected in subsequent
editions.

Cover and book design: David Barringer
Photo credits, page 320

For Carol, Lily and Maya

TABLE OF CONTENTS

Part I

Origins,
or
How Ball Brothers
Reached Space

Chapter 1

Rocket Projects

A **one-armed man sat in a concrete house.** Outside, repurposed German military hardware was about to blast his work into the great beyond. It was a typical afternoon in this corner of the American Southwest.

The house had walls twelve feet thick. Its roof was more substantial yet, solid man-made rock tapering into something approximating a pyramid. It had been designed to withstand a direct hit from a fully loaded V-2 rocket coming in at 2,000 mph.

And just up the way was a fully loaded V-2 rocket. On June 28, 1951, the adapted German terror weapon stood near its White Sands Proving Ground gantry, bloated with ethanol and liquid oxygen. The V-2 was topped with an experimental device called a biaxial pointing control. It looked like an ordinary rocket nose. It was much more.

Russell Nidey had helped design and build the pointing control. Nidey was a 26-year-old graduate student in physics at the University of Colorado. He was earning his master's degree for his contribution to the pointing control, a set of photoelectric eyes he had invented. The sensors would help the pointing control find the sun from the nose of a rocket soaring 50 miles up. If all went well, the pointing control could afford scientific instruments it carried the chance to capture certain bands of deep-ultraviolet sunlight that never reach the ground. Scientists and soldiers alike were interested in such light, which they believed might help solve astrophysical puzzles and improve military communications.

This particular V-2 was called Blossom IV-F. The series of V-2s known as Blossoms was so named because they had parachutes that would, in theory, soften landings for the contents of the rocket's noses. Those contents were no longer explosives, as had been the case with thousands of V-2s the German Luftwaffe rained upon Allied Europe during World War II, but rather tools of science.

Nidey was at home in high-desert heat. He grew up on the parched plains of southeastern Colorado, born in Campo, as close to Oklahoma as Colorado gets, and raised with five brothers and three sisters in nearby Springfield. In its sandy soils, his dad farmed broomcorn, a type of sorghum whose straw was prized by broom makers.

Nidey wiped his brow with a left short sleeve from which nothing extended. He'd lost the arm when he was three years old. He had fallen from a car, and his arm had been crushed by a rear wheel. When he was in high school, his mom had gone to the trouble of arranging for a prosthetic arm via a state social-services agency. The arm was wooden and ornamental, and, as Nidey saw it, in his way. During a welding class, he accidentally set it on fire, and soon abandoned it. He learned to button his right sleeve with his teeth and to cut his own hair. He probably could thank his missing arm for his presence in this blockhouse. He alone among his brothers had earned a high-school diploma, and he was the only one of his siblings to attend college. His parents knew he couldn't make a living with his hands.

Nidey was among 40 people, most of them graduate students, on the University of Colorado team. They had spent frantic months designing, building and testing to prepare for this 52nd American launch of a V-2 rocket. Nidey and a handful of students had then driven thirty hours from Boulder to southern New Mexico in an Army surplus six-by-six truck pulling a converted Air Force surplus photographic trailer.[1]

The members of the Field Operations Group, as the University of Colorado visitors to White Sands called themselves, had then spent days running final tests on their biaxial pointing control, and had kept to themselves their worries that it might not work. They had watched as Army personnel mounted their creation, a rocket tip just 15 inches wide at its base, to the much more rotund V-2's body using a makeshift adapter. They had examined the various buttons, switches and dials of the V-2 firing desk in this fluorescent-lit blockhouse bunker—gauges explaining such things as the oxygen pump inlet pressure and combustion pressure, and, on the sloping front panel, parallel rows of switches and jeweled lights with labels reading Master Power, Heater, Fuse, Pressurize, and, tantalizingly, *Fire*.[2] They had considered that, had they been here a year earlier, they might have witnessed German rocket pioneer Wernher von Braun himself prepare their V-2 for flight. But von Braun and most of his team had since moved on to Alabama's Redstone Arsenal.

A V-2 launch began with a yawning roar as the beast awoke, Nidey had been told. Then there would be an odd silence as the rocket raced away at speeds approaching a mile per second en route to a peak altitude of perhaps 90 miles. The only proof of its existence would be the contrail parting the blue. Then a few minutes later, its mission complete, various pieces of the rocket would plunge back to Earth somewhere amid roughly 3,200 square miles of White Sands desert. The Army's V-2 people made a point of detonating the rockets into sections on the way down to avoid the 30-foot-deep craters fully formed V-2s dug in the mesquite, cactus and creosote-dotted gypsum when landing intact at 2,000 mph. Nidey was looking forward to all this.

It was about a quarter to three in the afternoon. Outside, the sun beat down. Nidey looked out through a four-inch-thick slit of glass at the V-2 about 500 feet away. It carried nine experiments. One experiment measured the intensity of solar radiation at the far reaches of the atmosphere. Another captured X-rays; another infrared light; another sought to measure the brightness of the sky. There were also three movie cameras and a high-speed still camera on board.[3] Researchers from the Air Force, Rhode Island State University, the University of Utah and the University of Denver had many months of toil riding on this V-2. On top of the rocket perched the dart tip that was the University of Colorado team's product. It made this V-2 top out at about 50 feet, three-and-a-half feet taller than most.[4] An instrument inside the pointing control, built by Nidey's colleagues, would try to see deeper into the sun's ultraviolet than ever before. Nidey's photoelectric eyes were critical to achieving this feat.

Army men wearing headsets spoke in acronyms and numerical strings, confirming the rocket's readiness and making sure the men at various radio, radar and optical tracking stations dotting the ridges around the range were at the ready. The countdown began, backwards from twenty, ending with the word *"Fire!"*

Through the thick glass, Nidey watched as smoke spread from the V-2. At about that moment, the operator pressed a button labeled "Main Stage." Two powerful pumps began pushing 275 pounds of fuel per second into the fire. The roar intensified as the rocket rose.

It looked like a textbook V-2 launch for about eight seconds. Then an explosion blew the tail fins off. The rocket seemed unfazed, disappearing from sight.

The hurried chatter of the Army men confirmed the impression that things were going askew, a sense underscored moments later when tense voices began barking: "Abort! Abort!"

Up there, the V-2 motor went silent. Small TNT charges blasted the rocket's warhead from its body. The tumbling parts peaked at about three-and-a-half miles altitude. Nidey had been banking on at least 47 miles higher for his experiment to have a chance. The rocket sections then began a 45-second tumble ending in the bleached desert about a half mile from the launch pad.

The parachutes (there had been two, one for the lower portion of the warhead and a second for the biaxial pointing control) never opened. The work of Nidey and his University of Colorado colleagues was demolished. His efforts—and those of all the other scientists, engineers and technicians who had worked so hard for this day—had been for naught.

To UNDERSTAND what Russ Nidey and his University of Colorado colleagues have to do with the story of how a jar company hit a comet, one must begin with the V-2. The *Vergeltungswaffe zwei*, or Vengeance Weapon 2, was the world's first ballistic missile. The V-2's engine combined potato ethanol and liquid oxygen to propel the rocket to an altitude of 50 miles before burning out and landing a ton of explosives 180 miles away. Nazi Germany launched more than 3,200 V-2 rockets, most of them aimed at Antwerp and London. From September 1944 to the war's end in May 1945, V-2s killed an estimated 5,400 people in Europe, half of them in greater London. That doesn't include the thousands of slave laborers who perished building V-2s at the German rocket-development center at Peenemünde on the Baltic coast and in the dank, pestilent tunnels of the Mittelbau-Dora concentration camp near Nordhausen in central Germany.

Born of human suffering, the V-2 spent its days terrorizing civilians, both by design and because it was so inaccurate that it could find a target the size of a military installation only serendipitously. The Brits and Belgians would not be cowed, and the fearsome new technology would fail to change the course of the war. Still, despite its technical shortcomings, its dark intent and its Holocaust heritage, the V-2 counts among the world's great engineering achievements.

The V-2 rocket was the product of a 13-year development effort costing the equivalent of more than $15 billion in today's dollars. The V-2's lack of strategic return for the Reichsmark tainted it as another of Hitler's many miscalculations. But the Germans solved problems that had stumped rocketeers in the United States and elsewhere, such as how to keep a rocket's own heat from melting it.[5] Plus, the V-2 had made "complete nonsense out of strategic frontiers," as CBS newsman Edward R. Murrow observed in a November 1944 dispatch from London.[6]

Both the American and Soviet militaries considered V-2s war booty of the highest order. In the war's waning days, the U.S. military made off with several complete missiles and hundreds of rail cars filled with V-2 rocket parts from what would soon be Soviet-controlled East Germany. The Americans also brought home about 120 German scientists and engineers central to the program.

Foremost among them was Wernher von Braun, the top engineer for the world's first industrial-strength rocket program. Von Braun had been a teenage amateur rocket enthusiast with dreams of human spaceflight. The German military tapped him to help develop missiles in 1932. In 1945, von Braun and many colleagues wanted to complete their unfinished masterpiece, and they considered the United States their most viable patron. "We despise the French; we are mortally afraid of the Soviets; we do not believe the British can afford us, so that leaves the Americans," one of von Braun's men explained.[7]

Rather than spearheading a 1,000-year Reich, much of the remaining V-2 fleet ended up in some sun-scorched warehouses at White Sands Proving Ground in southern New Mexico, where the U.S. military was eager to test its new German hardware. The German rocket experts themselves, led by von Braun, learned English with a Texas twang at nearby Fort Bliss in El Paso.

VON BRAUN and colleagues led the technical effort of rebuilding the V-2s. Americans wanted working rockets for strategic reasons. But even in wartime Germany, von Braun and others had recognized the scientific value the V-2 might deliver. In 1942, he had commissioned German physicist Erich Regener to outfit a V-2 with instruments to measure how the atmosphere's temperature, pressure and density changed at heights tens of miles higher than any balloon had flown.[8] The approaching Soviet army had cut the effort short.

American scientists would pick up where Regener left off. With the Germans' help, V-2s would become the world's first sounding rockets, which leap into the sky, make scientific observations, and fall back to Earth not far from where they left.

"By doing this we showed that the big rocket, just like the big aeroplane, could carry more useful loads than bombs," von Braun later wrote.[9] And so the world's first space science happened on German missiles launched from an American desert.

The science still served military interests. A better sense of the atmosphere's various layers would help engineers sharpen the accuracy of a projectile plunging through them. Similarly, understanding the ionosphere—a 200-mile-thick layer starting at about 50 miles in altitude, known to influence radio transmissions—was central to commanding and controlling rockets of all kinds, be they scientific missiles or ones carrying bombs.

The ionosphere, which bounced certain radio waves back to Earth, enabled conversations around the planet in the days before communications satellites. Radiation showers from solar storms doused long-range radio talk with static. Scientists suspected blasts of short-wavelength ultraviolet light from the sun to be the culprit, but until the V-2, they had no way of proving it.

The problem was the stratospheric ozone layer. The ozone layer, a sheath of thin air between 10 and 20 miles up, protects the planet from all but 0.4 percent of the sun's ultraviolet light. Even this tiny dose suffices to mete out sunburn, skin cancer and cataracts.[10]

Most solar energy lands here as visible light. But the sun reveals itself most honestly in wavelengths blocked by Earth's upper atmosphere—a dirty window, in effect, through which scientists squinted at the heavens. Scientists resented the ozone layer, bemoaning the very ultraviolet-blocking traits that made it so vital to life on Earth. Without a better picture of the ultraviolet, they had scant insight into the makeup of stars and thus a shaky foundation for theories of how stellar bodies live and die—and, by extension, how the universe may have evolved. Without the knowledge of how much and what stripes of ultraviolet light were arriving at Earth, one also couldn't say what the sun did to the atmosphere or how, precisely, it affected people and the technology increasingly central to modern life, military or civilian.

The Nazi V-2, the first human creation ever to surmount the ozone layer, could help answer some of these questions, scientists hoped.

The rocket quickly delivered. An instrument on the 12th U.S.–launched V-2 would be the first to capture the sun's ultraviolet from beyond the ozone shield. In October 1946, the rocket climbed 90 miles and plunged back with a U.S. Naval Research Laboratory team's film exposed into the deep purple. The *Washington Post* published an image of the captured spectra on its front page; *Sky & Telescope* magazine called it "an event of far-reaching astrophysical consequences."[11]

But the German creation had fatal flaws. Clyde Tombaugh, the New Mexico astronomer who had discovered Pluto in 1930, bolted camera telescopes he invented onto a World War II M45 antiaircraft gun platform and tracked V-2s soaring as far as 100 miles away. As the rockets flew into the vacuum of space, with their fuel exhausted, their fins became as suddenly vestigial as those of a fish tossed in the air. The German rockets gyrated wildly—even tumbled through the heavens, Tombaugh observed—a dizzying ride for instruments trying to aim at the sun.[12] The V-2 could only carry science so far.

•

THE AEROBEE was the V-2's primary scientific successor. It was named after the rocket builder Aerojet Corporation and the U.S. Navy-financed Bumblebee program that paid for it. Just 15 inches in diameter and 26 feet tall, the Aerobee was an archer's arrow compared to the buxom V-2, which was 20 feet taller and five-and-a-half feet thick at its midriff. The first Aerobees could lift about one-tenth of a V-2 payload to an altitude of about 75 miles. But Aerobees were inexpensive and required neither German parts nor expatriates to keep them flying. The rockets were also spin-stabilized in flight, rotating up to twice a second as they soared. The effect provided an aerodynamic advantage over the V-2 similar to how bullets fired from a rifle are more accurate than balls shot from a smoothbore musket.

Aerobees presented their own challenges. If the rocket was spinning, the scientific instruments riding it would be, too. Any instrument hoping to collect data from a certain spot in the sky—the sun, say—would need to be somehow counterspun and steadied. Without new pointing technology, the Aerobee's advantages would be scant.

The first Aerobee flew in November 1947. Early launches carried cosmic-ray and radio-frequency experiments undisturbed by the rotation. But the U.S. Air Force was searching for experts who might build them something capable of aiming an instrument at the sun from a spinning rocket. The University of Colorado was an unlikely candidate—probably on only one man's list at all. He was Henry A. Miley, a physicist with the Air Force Cambridge Research Laboratories in Bedford, Massachusetts. In 1929, Miley had earned the second physics PhD ever conferred by the university. He saw the Aerobee problem as one suited to his former advisor, William B. Pietenpol.[13]

Pietenpol, 62, was chairman of the University of Colorado department of physics. A tall, soft-spoken man, Pietenpol was immersed in campus life,

serving as a timer at track meets and playing tennis with Fred Folsom, the law professor and coach after whom the football stadium would one day be named. The physicist was best known for his showmanship in the classroom. A former graduate assistant referred to the department chairman's lectures as "Pietenpol's magic hour." To explain why a sheet of paper fluttered rather than plummeted to the ground, the professor tore a page from a notebook, released it, and turned to the chalkboard to write an equation underscoring the point. To gasps and laughter, the paper fell like a stone. Pietenpol had loaded it with lead wire.

William B. Pietenpol

Pietenpol kicked off his lecture on frequency with a childhood story about how one of his chores was to chop firewood for the kitchen stove. He then produced an armload of sticks and tossed them into a concrete sink one after the other. They clanked out the tune of "Home Sweet Home" or "Yankee Doodle," depending how Pietenpol flung his whittled wood.

He displayed what students assumed to be a preternatural gift for drawing flawless no-look chalkboard circles with a great swing of his arm behind his back. One night, a student noted lights burning in the physics lecture hall and peeked through the door. There stood Pietenpol, drawing one circle after another, turning after each to check his form.[14] It was not surprising that Pietenpol, devoting such focus to the art of education, had "no strong research and publication history," as one chronicler put it.[15]

Indeed, the CU physics department was unremarkable. "Research and publication had not been absent in the preceding decades, but they had not been the dominant focus of the department, in part because there had been a near total absence of funds for the support of research in Colorado," as the department history put it.[16] Its core professors were approaching retirement. One student joked that the CU physics department was "the only one in which the optics man is blind, the acoustics man is deaf and you can't get a spark out of the electricity man if you tried all day." Another professor seemed "much more interested in the cattle that he owned than with what he was teaching."[17]

Complicating matters was a student body having little in common with the fresh high-school graduates to whom the aging professors were accustomed. Many students had returned from wartime service motivated to earn their degrees and get to work. They were impatient, driven, and largely unawed by academia.

Miley came to Boulder and met with Pietenpol in November 1947. Intrigued, Pietenpol assembled a team of professors and graduate students to brainstorm ways to help the Air Force point rocket-mounted instruments at the sun.

Key to the effort was the "deaf" professor, Frank Walz, who was indeed quite hard of hearing. He understood servomechanical control systems, which would be central to any attempt to steady instruments on a soaring rocket.

The CU team proposed to use sensors to spot the sun, then employ electric motors to spin the rocket's nose cone in perfect opposition to the Aerobee's whirling, thereby canceling out the rotation. As the rocket reached its apex, a mechanical arm would jackknife out of the rocket nose. The arm would aim the instrument at the sun like a cowboy pointing his pistol from atop a galloping horse, compensating for the rocket's precession—the wobbling of a football thrown slightly out of spiral. Servo-control systems would be at the heart of both the counterspin and the mechanical-arm mechanisms.

The CU team called their proposed creation a "biaxial pointing control," a clinically descriptive if uninspiring name for what would, if successful, become the first creation capable of pointing accurately in the vacuum of space.

The Air Force liked it. On April 13, 1948, the *Denver Post* ran a story with the headline, "Air Force Gives C.U. $69,000 for Research in V-2 Study Link."

It was the largest scientific grant in the university's history, and it involved much more than pointing at the sun. The CU team also promised to design an "air-borne coronagraph," which, in addition to observing the sun's faint atmosphere, or corona, would investigate "temperature, pressure, air composition, nature of the ionosphere, sky brightness, solar spectra, electric and magnetic effects and other highly technical aspects of the atmosphere 100 miles above the surface of the earth," as the newspaper story put it.

Given that no rocket-borne device had ever pointed at anything—and that the Naval Research Laboratory and the esteemed Johns Hopkins University Applied Physics Laboratory had tried and failed to produce something similar—the Colorado effort was hugely ambitious even without the coronagraph.[18]

Pietenpol and Walz brought in two PhD students to run the hands-on effort, James Jackson and David Stacey. The group became known as the "Rocket Project." They consulted some of the greatest minds in solar and rocket-based science. The day after the news of their Air Force grant broke, Pietenpol, Walz, Jackson and Stacey visited White Sands Proving Ground to observe a Naval Research Laboratory team led by Homer Newell as it prepared for a V-2 launch of a "Sunseeking Spectrograph." The Navy group "went out of their way to inform our men about the project," Pietenpol reported. The relationships they established with Newell and his Naval Research Laboratory colleagues would pay tremendous dividends.[19]

A month after the visit, members of the group visited James Van Allen's Applied Physics Laboratory team as it prepared for a V-2 solar-science launch at White Sands. Van Allen shared a laundry list of rocket pathologies with the CU team, from their irritability in flight to their seemingly unquenchable desire to pulverize scientific instruments on the desert floor.[20]

The CU group consulted with the famed solar scientist Walter Orr Roberts of the High Altitude Observatory, which had moved its offices from Climax, Colorado, to the gentler climate of Boulder. As a Harvard College graduate student, Roberts had built North America's first coronagraph on a mountain near Climax in 1940. His discovery of the link between solar flares and radio

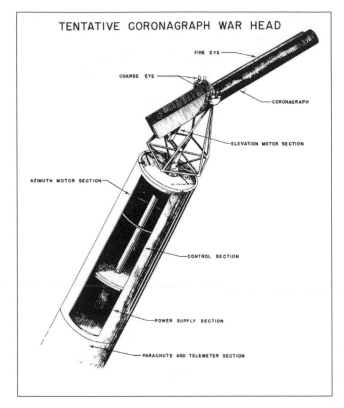

The "rocket-borne coronagraph" as first conceived

interference kept him there through World War II, his observations vital to the U.S. military's maintaining uninterrupted communications.[21]

Two of Roberts's High Altitude Observatory colleagues worked informally with Pietenpol's group. One was John W. Evans, who would become the founding director of the National Solar Observatory at Sacramento Peak, New Mexico; the other was Yngve Öhman, later to become the "'grand old man' and first pioneer" of Swedish solar science.[22] They focused on the subtleties of solar astronomy and how one might go about building their rocket-borne coronagraph.

Conversations with such brilliant men, in addition to design work exposing the true nature of the challenge ahead, quickly brought the Rocket Project back to Earth. Pietenpol cautioned his Air Force benefactors: "The University of Colorado group is fully aware of the difficulties involved in the research program outlined, and that the time and money required may be considerably more than allocated in the present contract."[23]

Pietenpol's group realized that they had promised the Air Force the impossible. The tip of an index finger at arm's length covers about one degree of sky. The sun itself is a half-degree across as seen from Earth. CU's flying coronagraph would need pointing accuracy of an arc minute, or one-sixtieth of a degree—roughly the size of a bottle cap as viewed from a football field away. The pointing control holding the coronagraph would have to shake off the abuse of an Aerobee launch, then aim with arc-minute precision despite rolling and lurching as the rocket soared above the atmosphere. The device would have to shoot bull's eyes from the back of a bucking bronco—and with a gun that existed only in theory.

Even if the CU team pulled off a miracle of pointing accuracy, there was the matter of the rocket-borne coronagraph. The instrument's lenses and mirrors had to be smaller than those in ground-based solar observatories. Smaller optics meant less tolerance for flaws. Such an instrument would demand something close to optical perfection, which was far beyond the technology of the day.

Pietenpol had proposed to invent fantastically advanced technology for a relatively modest sum, with delivery in little more than a year. Months melted away as a Rocket Project team of 20 and growing burned through their contract. There was progress: a working prototype of the system that would spot and lock onto the sun, which the team simulated using a 500-watt bulb in a film projector; a first-of-its-kind microscope to spot tiny imperfections in coronagraph lenses (it found many); even a "bench model" coronagraph in a wooden case. Still, the CU group was failing.

The Air Force could have sent Pietenpol and his student team packing. But Miley and his military colleagues sensed how difficult success would be. Space science demanded mastery of an environment entirely foreign to human experience. It involved speeds and forces we were just becoming familiar with. Exploring space was going to cost more than anyone had imagined—and demand of its patrons the patience of Job.

With the Air Force's blessing, the Rocket Project quietly reshuffled. Rather than develop the coronagraph, the CU team would develop a "monochromatic camera," later known as a spectroheliograph, to photograph the sun in a single ultraviolet wavelength. To help interpret the image the camera brought home, the Rocket Project would fly a sidekick ultraviolet spectrograph in the same pointing control. The spectrograph would arrange the captured electromagnetic jumble of sunlight into a sort of piano keyboard of ingredient wavelengths, allowing scientists to see what chords the sun plays in the deep ultraviolet light that never reaches the ground. The Rocket Project would soon hang its fortunes solely on the spectrograph, despite going to the trouble of building the much more complicated monochromatic camera. Like the scrapped coronagraph, the camera, too, demanded better accuracy than their sun seeker could hope to deliver.[24]

Rather than images and spectra of the sun's corona, the Rocket Project's target was now an ultraviolet band of light called Lyman alpha. It's the light a hydrogen atom emits when its electron, having been jolted briefly to the first energy level of the atom's electron shell, falls back to its unexcited ground state. Scientists believed such electronic action should be common in stars, which are full of both hydrogen atoms and energy to excite electrons. The Naval Research Laboratory's ballyhooed V-2 ultraviolet experiment of 1946 had reached only about halfway to Lyman alpha. It would take revolutionary technology on a different kind of rocket to plumb the ultraviolet depths.

For the Rocket Project, capturing Lyman alpha amounted to more than a $69,000 question. The team had won a reprieve after failing to deliver on the fanciful. Building their biaxial pointing control and detecting this coveted band of light—a merely improbable feat—could put the University of Colorado on the scientific map.

Chapter 2

The Sun Seeker

The University of Colorado physics department's Upper Air Laboratory, as the Rocket Project came to be officially known, consumed the basement of the Hale Scientific Building. Hale, with its Lyons sandstone walls and gothic roof, was among the university's oldest structures. In 1894, the physics department had moved across the green to Hale from Old Main, the university's first building, which for years had served as the higher-education equivalent of a one-room schoolhouse. The Air Force work displaced classrooms, offices and labs, some of which ended up in barracks-style military-surplus structures trucked to the lawn outside.

For evicted graduate students, a timber mezzanine was built over the main pointing-control engineering and assembly area. They worked through noise from the sun-seeker factory below as well as other distractions. "Unfortunately the plumbing wasn't so good," a displaced student remembered. "We had a circle marked on the floor, and so whenever we'd hear the flush upstairs, I'd have to move my chair forward because there'd be a few drops of water come down and drip on the floor on that spot. We thought it was entertaining at the time."[1]

The Rocket Project had a main laboratory as well as a design room. Sitting at drafting boards, graduate students rendered sketches of various components and systems, each line and curve justified by calculations made with abacuses and slide rules. There was a machine shop, designed for building or refining components and hardware, and two small offices. One was home to two secretaries; in the second worked the physics PhD students James Jackson and David Stacey.

Jackson and Stacey ran the Rocket Project's day-to-day operations. Jackson focused on project management, Stacey on the technology. Jackson was a stocky five foot six with round glasses and slicked-back hair. He wore jeans with cuffs rolled up a good three inches and smoked a pipe. Jackson hailed from L.A., the tongue-in-cheek name for the farming community of Las Animas in southeastern Colorado.

James Jackson, left, William Pietenpol, and David Stacey, kneeling, with a biaxial pointing control outside the Hale Physics building

Jackson's grandfather had owned a ranch. His father had been a civilian doctor at the Navy tuberculosis hospital in nearby Fort Lyon. The family moved to Battle Creek, Michigan, when Jackson was young, and he graduated from Michigan State with a degree in physics. He worked on motor efficiency at the Armour Research Foundation in Chicago for a year, spent a year back at the family ranch, and arrived at CU in 1946.

Just 25, Jackson liked to be in charge and was a natural administrator and workhorse. His energy and directness, as one graduate student remembered, made him "not particularly well-liked" by the physics faculty. "He was a bit aggressive in accomplishing what he wanted. He didn't mind stepping on other peoples' toes at all if that served his purpose. But on the other hand, he didn't run roughshod."[2]

Stacey, 30 when the project began, was six inches taller and more interested in building things than leadership. He was happy to let Jackson handle administrative matters.

Stacey was born in Cincinnati, Ohio, and had graduated from Harvard in 1940. He spent the early war years as a glider flight instructor in New York and California. By 1942, he was working on underwater demolition at the Woods Hole Oceanographic Institute in Massachusetts. Stacey, like Jackson, started his PhD work at CU in 1946. Stacey's avocations involved mechanical tinkering and physical adventure, including hiking, kayaking and piloting one of two gliders he owned. Boulder was a perfect home.

His middle name, Stearns, came from his mother's family. She was a Stearns of Stearns & Foster, the Ohio maker of mattresses, and she died when Stacey was a boy. Family money put him and his brother through private

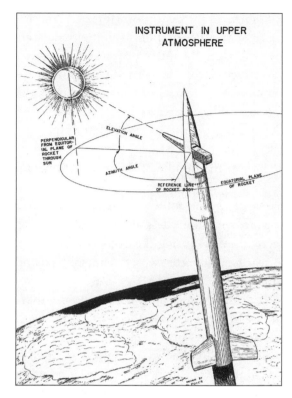

INSTRUMENT IN UPPER ATMOSPHERE

Artist's conception of a biaxial pointing control in action

school in Colorado Springs and eventually bought the Boulder home across the street from the city's signature Flatirons rock formations, where Stacey, his wife Joan, and their two infant sons lived. The source of his wealth would remain mysterious to his colleagues. Stacey's lack of airs made it little more than an occasional source of speculation. He was a laconic man whose words, when he chose to speak, carried weight.

Jackson and Stacey both considered the Rocket Project's success more important than their own academic progress. It took Stacey eight years to earn his PhD; Jackson never would get his.

Stacey's 1954 PhD dissertation, 47 pages of mathematical equations with occasional English interruptions, was a theoretical piece arguing for the superiority of something he called the "Ideal Control." A type of servomechanism, the Ideal Control was the Rocket Project's intellectual foundation.

A household thermostat is a servomechanism—something that, as Stacey explained it, "compares a desired quantity with an output quantity and then attempts to reduce the difference to zero." In the thermostat's case, a controller detects a room's actual temperature (the *output quantity*) and compares it to the *desired* temperature indicated on the dial (which varies depending on whether the homeowner wears sweaters or T-shirts in the depths of winter). A "motive power" section—in this case, a furnace or an air-conditioning unit—then alters the room temperature in an attempt to reduce the difference to zero.

A typical servomechanism moves precisely to the desired target and stops in a linear way. Linear servos limit their speed so as to halt smoothly at the proper spot, like a driver at a stop sign. Stacey's servo, in contrast, cranked its electric motors full-bore so the pointing control zipped slightly past its target, then reversed to swing back to that target. Stacey's servo was a driver who gunned it, slammed on his brakes in the middle of the intersection, and smoked the tires backing up to the stop sign.

A thermostat wired with Stacey's Ideal Control would alternatively stoke the furnace and blast the air conditioner in an endless, energy-wasting duel. But similar servos, developed in secret at MIT during the Second World War, had been vital in directing the guns of navy battleships, where split-second targeting delays could get a boat sunk.[3]

On a Rocket Project pointing control, photoelectric sensors set the thermostat, so to speak, finding the sun for Stacey's servomotors. Detecting "eyes," as the team called them, translated the sun's rays into an electrical charge.

Russ Nidey designed the optical system. Nidey's system involved 21 such eyes of two varieties—"coarse" and "fine"—arrayed about the biaxial pointing control. The electrical charge sent from a given eye varied with the brightness of the sunlight striking it. Nidey orchestrated the perception of all 21 eyes to enable the pointing control to find its target, despite the unpredictable flickering of light and darkness during the spinning pointing control's flight.

Next came the challenge of building and packaging pointing control hardware, scientific instruments, and various electrical, communications and other systems into a roughly eight-foot-tall rocket nose just 15 inches wide at its base. In this oversized dart tip, Rocket Project hardware would have to survive the acceleration and vibration of launch, temperatures as high as 500 degrees on its ascent, and, probably, crash landings in the New Mexico desert. The team had to mind weight, as each additional pound of payload would cost 1,000 feet of altitude.

The Rocket Project spawned separate labs for optics, general mechanical work, servomechanisms and electronics. Student designers penciled at their drafting boards as professional machinists in the shop created parts on presses and lathes. Cheesecake calendars of topless women, taped to shop walls, interrupted a décor of cement, brick, wood and metal. Jackson and Stacey intended to build an outdoor heliostat, a sun-tracking device capable of reflecting rays into the test labs inside Hale. University administrators balked at the idea of

Rocket Project students Merle Reisbeck, left, and Russell Nidey look on as a colleague works on a biaxial pointing control.

cutting down trees planted, watered and nurtured to grow on what had been arid prairie. The team instead converted a surplus Air Force mobile photographic laboratory trailer into an optics testing lab into which they could reflect sunlight. As their pointing controls took shape, students rolled them outside and tested them on the lawn.

The actual pointing controls were only part of the effort. There was a sunseeker prototype, resembling a half-built android tangled with wiring, bolted to a cement floor, staring at a 500-watt faux sun. Its stacked control electronics were as tall as a man. Prototypes were vital to Stacey's work. He and colleagues realized the theoretical Ideal Control—touted in early Rocket Project reports to the Air Force as nothing less than a "Perfect Brain"—was "child's play compared to the problem" of aiming at the sun from the nose of an unsteady rocket.[4]

The CU students' mathematical rigor had been crucial, though. For one, it set the Rocket Project apart from earlier, unsuccessful attempts by the Naval Research Laboratory and the Applied Physics Laboratory. The numerical proofs blazed a trail for later hands-on experimentation while showing the Air

Force that the students' pointing controls *should* work long before the CU team could actually prove it in the skies over New Mexico.[5]

But Stacey recognized that mathematical manipulation could only carry the Rocket Project so far. In a paper-and-pencil world, running the same "program" with different inputs meant updating every calculation by hand. The possible inputs seemed endless; the team's window of time for theorizing was closing fast as months passed and the Air Force outlays mounted. The Perfect Brain proved little more than an imperfect baseline. Progress now would follow on a rocky path of trial and error.

The sun-seeker prototype

Developing a pointing control of any real precision, Stacey and his colleagues knew, meant piecing together a system that could *sort of* point at the sun through the chaos of rocket flight, and then polishing every facet of that system. The photoelectric eyes could not see with infinite sharpness. Parts had been machined by men, made of metal forged by men, on equipment made by men—fallible men asked to make parts accurate to within a ten-thousandths of an inch, or about a twentieth of a hairsbreadth.[6] Each meeting of gear teeth, each pulse of electricity flowing through the not-quite Perfect Brain, introduced variability, which undermined the men's ability to control the system.

REDUCING VARIABILITY meant designing and building novel support equipment. The students designed and built a gear tester, a spin-tester, and a g-force tester, which dropped things and then measured how hard they fell. They built their vibration tester, used for shaking pointing controls with the approximate vigor of an Aerobee launch, into a small trailer with a gasoline engine.

Otto E. "Pete" Bartoe, a 21-year old starting his master's degree in mechanical engineering, helped design the vibration tester. It was among the earliest of an amazing array of hardware he would conceive during his long career in Boulder.

The son of an officer in the U.S. Marines, Bartoe called Coronado, California, home but had spent years growing up in the U.S. Virgin Islands, Hawaii and elsewhere. As a 14-year old, he had observed flocks of Japanese Zeroes from his yard during the attack on Pearl Harbor. U.S. anti-aircraft rounds crashed back down around him, killing a woman in a house a few doors down.

Bartoe designed and built model airplanes on Oahu. He had skipped fourth grade and lettered in football and basketball before graduating at age 16 in 1943. He was powerfully built, over six feet tall, with short, curly hair and a scratchy voice an octave higher than one might expect. He spoke his mind; he did not suffer fools easily; he was "not capable of lying," as his wife Mary later put it. Bartoe spent a year at CU and then joined the Marines, where he trained until the war's end. He returned with a pilot's license and, toward the end of his studies, lived in a room at the Boulder Municipal Airport hangar. He and a friend slept in a ready room and served as line boys, gassing up planes and washing windows. Rent was free, and the sole utility cost was the electricity needed to power their room's bare blue 25-watt bulb.

Bartoe graduated from college in 1949, a time when, he said, "you could hardly get an engineering job in the United States. There were virtually none."

Grand visions of space exploration had nothing to do with his interest in the Rocket Project. It was his ticket to graduate school. "I figured if I can't get a job, I might as well get another degree," Bartoe said.

He went to work on a drafting board in the Hale basement. His first assignment was to design a mount for an instrument mirror the size of the bottom of a coffee cup. It had to move back and forth as well as up and down. "All I can remember was trying to figure out how the hell would you hold a mirror in the first place," Bartoe recalled. "What would you make such a mount out of? Physically, would you mill it? Would you saw it out?"

Bartoe, Nidey and their young colleagues faced problems new to engineering. William Pietenpol, while nominally leading Stacey, Jackson and the rest, limited his project duties to attending a weekly meeting and offering periodic advice. Physics professor Frank Walz was more hands-on in the servo laboratory. Dick Crawford, an engineering instructor, offered some guidance to Bartoe and other young design engineers. But this was a classic academic operation, with professors offering occasional suggestions to students doing most of the real work.

Surprises added to their workload. In 1949, the Air Force asked the Rocket Project to build an additional pointing control for an instrument a University of Denver group was building. This one would counterspin the rocket nose, but have no pointing arm.

It was an opportunity to test the eyes and motors of the control system needed to counterspin an Aerobee nose cone in flight, but it doubled the demands on a project still struggling to get its first biaxial pointing control to work.

Pietenpol agreed to build the University of Denver machine. The team, which between late 1948 and mid-1949 had doubled in size, leveled off at about 50 people for several months, during which students and machinists forged ahead on their original pointing control as well as the simpler, uniaxial version.

More than a year later, in October 1950, Nidey and a handful of other Rocket Project team members packed up their "warhead," as they called it, into a lumbering military-surplus truck and drove nearly 600 miles south to Holloman Air Force Base near Alamogordo, New Mexico. They spent more than a week in the desert preparing for the Aerobee launch of their first pointing control, the uniaxial version carrying the University of Denver payload. On November 2, 1950, the rocket shot off with a promising whoosh. But it ran into "high upper winds," which brought about an "extremely unstable flight with the missile precessing as well as rolling. Attitude was far from vertical and at times horizontal," as the official report put it.[7]

The Aerobee had flown like a snowflake in a gust. The pointing control's parachute never opened. The students gathered up their crunched hardware and drove back to Boulder.

The University of Denver uniaxial pointing control, before and after

Henry Miley of Air Force Cambridge paid his alma mater a visit. The biaxial pointing control prototype was slated for launch in spring 1951. That was still on, Miley said, but he wondered about the possibility of launching the CU team's second biaxial pointing control a couple of months later, in June 1951.

They would have precious little time to build and test another device, much less incorporate lessons learned from the biaxial prototype's maiden flight. What's more, the ride in June would be on a V-2 rocket.[8]

There was no reason why a biaxial pointing control couldn't work on a well-behaved V-2, if such existed. The issues were girth—the V-2 was much thicker and would require an adapter to host the svelte Aerobee pointing control—and the endurance of a team heavy on overtaxed students who also had classes to attend.

Pietenpol told Miley the added burden would mean all work to improve the engineering on the biaxial pointing control would stop, and that the project would hire extra people and operate "on an emergency basis."[9]

The basement of the University of Colorado's department of physics became a frenzy of design, machining and construction. The machine shop swelled to 10 craftsmen cranking out and fine-tuning parts for two pointing controls.

Piece by piece, the sun-followers' components arranged themselves in sparse elegance: gears the size of hors d'oeuvre plates, tributaries of wiring bound into orderly channels, black gear boxes, batteries. Missing were integrated circuits—computer chips—which had yet to be invented.

The pointing control's intelligence expressed itself not in the stark ones and zeroes of digital computing, but in analog shades. Photodetectors perceived the sun's brightness and sent voltage through a system of wires, amplifiers and junctions arranged to translate a bewildering set of possible rocket actions into immediate, precise counteractions.

At least that was the idea. There were still all sorts of problems. The nose-cone structure, to be sealed and pumped to a near-vacuum prior to launch to match its operating environment 50 miles above, leaked. The instrument arm, badly out of balance, required lead counterweights. Molybdenum lubricant splattered about, shorting out servo electronics.[10]

In the spring of 1951, the CU team was back in New Mexico, this time with its first biaxial pointing control. On April 12, it shot into the desert sky with the haste of a bottle rocket, vanishing abruptly from sight. But this Aerobee would fall short. A fuel line on the rocket broke 30 seconds into the flight, stalling its ascent at 21 miles.[11] The pointing control's parachute again stayed put, and the desert floor took no pity.

The first biaxial pointing control as the search team found it at White Sands

Somehow, the instruments in the pointing control survived. The monochromatic camera brought back images of sunny haze which, while of no scientific use, gave the team in the Hale basement reason to hope that their pointing system had worked. But years had passed, and hundreds of thousands of dollars spent, with little to show for it besides twisted hardware and a few blurry frames of film. The student team satisfied its unfulfilled need for success in subtle ways. In the Rocket Project's "design priorities," in addition to reliability, low weight and low cost, "appearance" became a factor. Looks were important, Bartoe explained at the time, because "a well-designed and neatly fabricated unit is in general more reliable. Also, since the field of rocketry is prone to unforeseeable failures, appearance is of great value concerning confidence and morale."[12]

Failures happened, and rocket science advanced with their lessons. That's at least what the student engineers tried to convince themselves. Defeated again, they returned to Boulder, where work on the V-2 pointing control continued.

UNUSED VACATION DAYS piled up. Jackson was owed more than 50 days; William Lowrey, who ran the machine shop, was behind more than 70 days.[13] In June 1951, the launch team packed up again for New Mexico, their second biaxial pointing control still riddled with glitches. They lacked both the equipment and the time to fix them in the desert.

The failure, on June 28, 1951, of the V-2 known as Blossom IV-F spared Nidey and his colleagues any potential embarrassment.

Pietenpol extended the Rocket Project's Independence Day break until July 23 to spell his exhausted team. "The Laboratory is now recovering from the pressure and disorganization of two extremely rushed firings," Pietenpol wrote. "An exception is the instrument shop, which has a pointing control to manufacture for a December firing."[14]

The endeavor had drained Rocket Project coffers despite lavish budget growth. To the initial $69,000 contract had flowed another $69,000 in October 1948; $166,000 in 1949; $198,750 in 1950; and $80,000 in March 1951.[15] Despite the string of failures, their planned production volume had grown, too—the Air Force now expected 10 pointing controls once the prototypes began working. The Rocket Project was running behind by about two months, and time was money.[16]

In August 1951, the CU team's leaders sent a frank summary of their financial affairs to the Air Force. In addition to asking for more funds, the Rocket Project's members had come to recognize an enduring truth of the space business. "Needless to say, it was impossible to estimate the development cost of an instrument which had never been built. In fact, there existed a reasonable doubt that it could ever be built," they wrote.[17]

By then, the CU group could at least wow Air Force visitors with a pointing-control demo. It involved setting up a prototype nose cone out on the lawn. A rotating, gyrating platform simulated an Aerobee's rotation and gyration. The pointing arm "would usually stay pointed at the sun," as one team member remembered.[18]

In late April 1952, the team packed up again for New Mexico, where their pointing control was slated to fly on a new Aerobee-Hi rocket. On the morning of May 1, 1952, the rocket climbed 56 miles in a flawless arc. The student-built hardware finally had its moment in the sun.

This time, the pointing control failed. The culprit was probably a cracked photoelectric eye on the pointing arm, which then took no particular interest in the sun. The instrument inside never had a chance.[19] The CU team had been working for more than four years only to watch three of their creations felled by bad rockets and a fourth by their own blind eye.

William Rense

Another six months passed before their fifth attempt. The sole instrument this time was a grazing-incidence spectrograph, which was deemed to have the best shot at capturing the Lyman alpha ultraviolet band. Bartoe did the spectrograph's mechanical design, and the machine shop cut its finely grooved mirror glass in-house, as they had done for previous instruments. The scientific mind behind the Rocket Project's instrument work was a young physicist named William Rense. Rense, in his mid-thirties, had been on the faculty of Louisiana State University and had taught summer school at CU for years. Pietenpol had hired the soft-spoken scientist when it became clear in 1949 that the coronagraph would never fly.

The Rocket Project's launch team caravaned to New Mexico for the fifth time in early December 1952. Shortly after noon on December 12, their work vanished into the morning sky. Ninety seconds later, springs jettisoned doors on opposite flanks of the rocket's nose. The atmosphere lay almost entirely below; above, the stars shone, the sun the brightest among many distant relatives.

Several of the pointing control's "coarse" eyes, mounted at the base of the nose cone, lit. The eyes, sensing the sunlight on the rotating projectile, conversed with servomotors in the language of electric charge through homespun analog circuitry. The motors spun the rocket's nose in perfect opposition to the rocket's slow roll.

The boxy pointing arm then thrust out of the nose cone's casting. Rense's spectrograph rode inside, its film still waiting in darkness. Gracing the pointing arm was a necklace of five photoelectric eyes. Two were the same as the coarse eyes on the rocket's neck, designed to position the pointing arm to within a degree of the sun. The rest, the "fine" eyes, would take over for precision tuning.

The arm flailed for a moment until the coarse eyes locked on. Stacey's theoretical Perfect Brain, made wiser by years of experimentation, kicked into action, throwing the full weight of the servos at the sun, overshooting, then backing down in little more than a second. The fine eyes then brightened, making continued, minor adjustments to both the lever-arm and rotation servos. Two minutes and 18 seconds into the flight, Rense's spectrograph awoke to the sun's glare.

Fifty-five miles above the desert, the Aerobee flew nearly straight at the sun and was scarcely rotating. For 28 seconds, the pointing control aimed with an accuracy of a quarter-degree as Rense's film soaked in an unprecedented view of the star.

Gravity took a firmer hold. The instrument pulled back into the safety of the nose cone and the servos ceased their rotation. The Aerobee's tail blew off, sending the rocket's shaft and nose into an intentional tumble to slow their fall through the thickening atmosphere. At 20,000 feet, the pointing control separated and waited in vain for its parachute to bloom. The sun-follower thumped onto the desert floor eight minutes after it took off.

Inside the bent pointing control's instrument casing, Rense's film negative had been burned with a ghostly line where the spectrograph directed light with a wavelength of 121.6 billionths of a meter. The Rocket Project team had captured Lyman alpha.

A single, faint streak on a 16-inch stretch of film may have seemed a poor trade for four years of hard work and hundreds of thousands of dollars of tax-payer investment. But the U.S. government was willing to spend vast sums on hard data shedding light on even a sliver of creation, a habit that would build an industry.

Scientists took notice. James Van Allen, the physicist who had been so helpful at White Sands days after the group's creation in 1948, sent Pietenpol a letter, saying, "Please accept my congratulations for this classical achievement. This is certainly the most exciting news in the field in many years." Richard Tousey, the Naval Research Laboratory scientist whose V-2 ultraviolet spectrum had landed in the *Washington Post* in 1946, flew to Boulder and ordered several Rocket Project pointing controls for his own team's research.

When Rense published his findings early the next year, the national media took note. In February 1953, *Time Magazine* reported the achievement in a story titled "Sun-Seeker." It credited Air Force Cambridge and the University of Colorado with creating "a gadget specifically designed to do the job [of photographing the ultraviolet] from a high-flying rocket." The *New York Times* got around to the story eight months after the event. The Rocket Project's feat shared a column of "Notes on Science" with a helmet for supersonic pilots, a bomb to detect the location of sinking ships, and a rundown on the economics of flight insurance. Fame had been fleeting, and few understood the significance of what the group in Colorado had actually achieved. Lyman alpha got the attention, but the real magic had been in the pointing control.

The biaxial pointing control, designed by students in a University of Colorado basement, had become the first device ever to find its footing in the great vacuum above. Americans—not Russians, not Germans—had made a machine that could focus on a particular point in space *from space*, a basic requirement of nearly all spacecraft that would follow. The pointing control was the Western Hemisphere's first major home-grown space technology.

Yes, it had all cost ten times what anyone had expected. True, the ultraviolet instrument "possessed no interesting features" as Rense himself described it.[20] Yet the simple instrument and the elegant pointing control had delivered.

The Rocket Project had little time to enjoy the moment. Pietenpol's team owed the Air Force nine more pointing controls. The next one underscored just how fine the line between success and failure in the business of rocket science was. The team decided to add a touch of decorative flair to their rocket nose. They settled on a shade of purple Nidey had noticed on his sister MaryAnn's fingernails—an uncommon paint color at the time. Students baked some of the nail polish (which they renamed "high-temperature lacquer") and assumed it would, worst case, flake off harmlessly. But in the heat of ascent it melted, spattering violet droplets up and down the rocket and blinding several of the pointing control's sun-seeking eyes.[21]

The CU team simply built another one. The Rocket Project's biggest challenges were no longer experimental. The physics department's traditionalists had grown impatient with what appeared to be a manufacturing shop in their basement. The university's loss would be Ed Ball's gain.

Chapter 3

The Weighty Business of Modernization

Boulder, Colorado, where the high plains met the Rocky Mountains, was a small mining town from its founding in 1859 until World War II. Twenty-seven miles northwest of Denver, Boulder had grown to 19,999 people as of the 1950 census. The number was inflated by thousands of University of Colorado students. These academic nomads, included in the count for the first time that year, comprised a quarter of the total.

Boulder was a scenic gemstone. At the city's western edge, the fields of America's breadbasket and arid plains finally surrendered to ponderosa pines and Boulder's signature Flatirons formation, a series of 1,600-foot-tall, delta-shaped monoliths looming over the city's southwestern flank. The Flatirons' pink sandstone had eroded from ancestral Rockies nearly 300 million years ago, and a tectonic flourish had thrust them upward as the modern Rocky Mountains rose thousands of millennia later.

The gold and silver mines in the foothills to the north and west and the coal mines south and east of town played out in the years prior to the Second World War. The only Boulder company of note before the war had been Western States Cutlery, which made Boy Scout knives. With resource depletion threatening the city's viability, Boulder's leaders sought to lure businesses to town. For all its natural beauty, the city would need a bit of luck. The whims of geology had built Boulder; those of geography and politics would spark its rebirth.

The National Bureau of Standards, whose scientists had developed the world's first atomic clock and wrote rulebooks for everything from steel strength to electronics, was to establish a laboratory far from Washington, D.C. The government sought a small-town location with little radio noise, a university and a nearby transportation hub. Charlottesville, Virginia; Palo Alto, California; and Boulder were the top contenders.

The Boulder Chamber of Commerce enticed Washington bureaucrats with photos of the Flatirons and the city's fabulous mountain backdrop. But the small-town leaders had a sense of what really makes government officials tick. In 1950, a chamber cash drive raised $90,000—about $800,000 in today's dollars—in just two weeks. City leaders spent $63,000 of it to buy 217 acres of pasture south of town, which they promised to donate to the federal government for the lab.

The local money and the influence of Colorado's powerful U.S. Senate delegation tipped the scales. In September 1954, President Dwight D. Eisenhower dedicated the new NBS building. There, 450 scientists, engineers and support staff worked on everything from atomic timekeeping to radio-propagation experiments. Not a few had lamented leaving the nation's capitol for the "scientific Siberia" of Boulder, a place they considered far removed from civilization and on the "western frontier," as one historian recounted, despite the rumored existence of California.[1]

The effort to lure the federal laboratory also secured Boulder's foothold in the space business, albeit indirectly. Money left over from the chamber of commerce cash drive paid for a 20-acre patch of land east of town they called the Boulder Industrial Park. For years, the expanse west of the Humane Society building remained vacant. A startup weighing-device company, led by an entrepreneur named Ed Safford, would change that.

Ed Safford

SAFFORD WAS A BORN ACHIEVER, an Eagle Scout who earned an engineering degree from the University of Kansas in 1939. He spent the first decade of his career in sales for Beech Aircraft in Wichita, traveling the world selling airplanes. He then founded truck-trailer maker Spencer Safford Loadcraft in 1950. Recognizing the need to weigh truck payloads more accurately, Safford came upon the idea of building a new sort of scale, which he called a control cell. In 1953, he launched Control Cells Corporation.

Safford's scale calculated heft electronically rather than via the springs and balances of conventional scales. By avoiding such mechanisms, the weighing device itself could be "only a littler larger than a package of king-size cigarettes," as a newspaper account put it.[2]

Safford, then 36, had enlisted the Foundation for Industrial Research at the University of Wichita to help develop the device. They worked on it for two years. In early 1955, Safford had moved his fledgling company from Wichita to Boulder, a place he had visited and liked. The company sold scales to trucking firms as well as toll-road authorities.

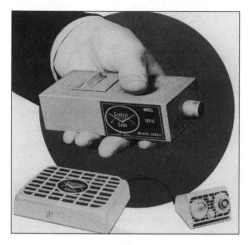

The Control Cell and one of the scales in which it operated

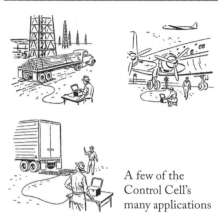

A few of the Control Cell's many applications

"Development and production of Control Cells provides manufacturers a revolutionary electronic weight-and-force-indicating device for an Age of Automation already here," declared a Control Cells brochure. The scales, either mounted directly to truck-axle assemblies or built into a "pancake" aluminum-alloy platform a vehicle could drive onto, could weigh a truck twice as fast as a conventional scale, the company promised.

But trucking was only the beginning. Safford believed his scales would measure batches of materials, fluids and gases; revolutionize inventory management in the beverage industry; and be at the heart of safety systems in elevators, cranes and oil derricks. Toll roads in Missouri and Minnesota were weighing trucks with the devices, and the Indiana Toll Road was installing them, too. United Airlines was testing the scales for use in checking aircraft loads at Denver's Stapleton Airport. Men with control cells determined that an 11-room, 2-story brick house in Boulder weighed 346,600 pounds.

"At Boulder, a company is making gadgets 5 inches long by 1 inch thick which are being used to weigh trucks, houses, bridges, railroad locomotives and heaven knows what else," the *Denver Post* reported.[3]

Behind the public-relations facade, Control Cells was struggling. That would change, Safford believed. In trucking alone, he anticipated installing four devices on each of 188,000 trucks by the decade's end—2 percent of the nation's fleet. That market alone would bring in $75 million a year by 1960, he told the Boulder newspaper.[4] Roughly $600 million today, it was a projection reminiscent of those touted by dot-com startups during the 1990s Internet bubble.

The 1950s too were an effervescent economic time. The Dow Jones Industrial Average had more than doubled from 1950 to 1955. A sense of optimism swept the land. "Once again the feel of a 'new era' is in the air," wrote *U.S. News & World Report* in May 1955. "Confidence is high, optimism almost universal, worry largely absent. War is receding as a threat. Peace is a growing prospect. Jobs are quite plentiful. Pay was never so good. The promise is that taxes will be cut. Everywhere things are in a rising trend."[5]

Companies old and new viewed technology as a key engine for growth. Few had more need to modernize than the Ball Brothers Company of Muncie, Indiana, maker of the famous glass jars. Its president, Ed Ball, was leading an effort to bring the old-line container company into the 20th century.

EDMUND F. BALL'S FATHER, Edmund B. Ball, and uncle, Frank Ball, had been farm boys. These two of five Ball brothers founded the Wooden Jacket Can Company in 1880 with $200 borrowed from a Baptist-minister uncle. The Buffalo, New York, startup's first big sellers were wooden-jacketed tin cans for storing kerosene destined for lamps.

The tendency of the acids in kerosene to corrode tin led the Ball brothers to glass. They quickly entered the burgeoning canning-jar business, which swarmed with competition sparked by the expiration, in 1879, of John L. Mason's patent on a molded glass jar with neck threads. Mason jars were a hot item for good reason. Before refrigerators and industrial food processing, home canning kept seasonal foods on the table through the depths of winter. The Ball Brothers Glass Manufacturing Co., so renamed in 1884, produced 1.8 million mason jars in 1886, in addition to jacketed cans and various bottles.

Along his travels as the company's principal salesman, Frank Ball received many offers to relocate. Muncie, Indiana, an agricultural county seat with grander ambitions, had recently enjoyed one of the Midwestern Gas Belt's biggest strikes. The town offered Ball $7,500, nine acres of land, a rail-line extension and free natural gas for five years.[6]

It was not quite love at first sight for the visitor from New York, who found Muncie's streets "dusty and very dirty," adding, "There was nothing about the town that particularly appealed to me, but the men were all courteous, kind and businesslike."[7] This was a business decision, based on the same cold

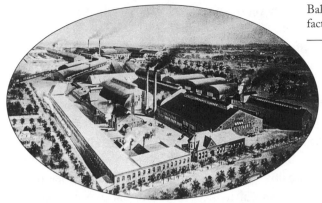

Ball Brothers' Muncie glass factory in the late 19th century

calculus that would one day uproot the Midwest's manufacturing base to Asian enterprise zones.

The company moved to the town of 6,000, producing its first containers there in 1888. The rest of the Ball family soon joined Frank and Edmund in Muncie. Home-canning jars became the Ball calling card, the company's strategy of ensuring a tight seal by producing both jars and zinc lids having proved to be a winner. (Competitors generally did either glass or metal, but not both.) Ball Brothers also constantly improved their production technology and pioneered such things as corrugated boxes to replace wooden shipping containers. In 1892, the company produced 9 million jars; in 1905, the year Ed Ball was born, they produced 74 million; in 1931, at the peak of the home-canning boom, it was 190 million jars.

As Ball became a household name and the Ball brothers grew immensely wealthy, they gave their money to Ball Memorial Hospital and what would become Ball State University. They built the Muncie Masonic Temple, complete with 900-seat auditorium. They touched so many aspects of life in their adopted hometown that the 1937 sociological study *Middletown in Transition* dedicated an entire chapter to the Ball family. Its title: "The X Family: A Pattern of Business-class control."

Said one "Middletown man":

> If I'm out of work I go to the X plant; if I need money I go to the X bank, and if they don't like me I don't get it; my children go the X college; when I get sick I go to the X hospital; I buy a building or a lot in the X subdivision; my wife goes downtown to buy clothes in the X department store; if my dog stays away he is put in the X pound; I buy X milk, I drink X beer, vote for X political parties and

get help from X charities; my boy goes to the X Y.M.C.A. and my girl to their Y.W.C.A.; I listen to the word of God in X-subsidized churches; if I'm a Mason I go to the X Masonic Temple; I read the news from the X morning newspaper; and, if I am rich enough, I travel via the X airport.[8]

The book went on to describe the surviving Ball patriarchs as "alert, capable, democratic, Christian gentlemen, trained in the school of rugged individualism, patrons of art, education, religion and a long list of philanthropies; men who have never spared themselves in business or civic affairs; high exemplars of the successful, responsible manipulators of the American formulas of business enterprise."[9] Despite its wealth and power, the Ball family had many fans in Muncie. The year *Middletown in Transition* was published, 11,000 local donors contributed to pay for the statue "Beneficence"—an angel with a jewel box in one hand and the other outstretched in offering—in honor of the five Ball brothers.[10]

Ed Ball grew up in a limestone mansion on Muncie's White River. The grand homes of the five Ball brothers graced the same tree-shaded street. The homes were arranged by rank, Frank Ball's estate being the capstone farthest east. Ed Ball attended private schools and graduated from Yale with a degree in philosophy in 1928. His yearbook photo depicts a young man handsome in a soft-featured, clerical way, with oval glasses rounding out the impression. The 23-year-old had fancied going on to graduate school, but his uncles Frank and George—his father had died of a stroke three years before—had other ideas. Ed came back to Muncie to work on the factory floor.

As he climbed the company ranks, Ball fell in love with flying, going so far as buying the Muncie airport with his cousin Frank, son of the family leader and heir apparent to the Ball Brothers throne. But the young Frank Ball died in 1936, after the wing of a biplane he was flying collapsed over Findlay, Ohio. Ed Ball became logical successor to Ball's aging family leadership.[11]

In 1931, he had joined the Army reserves. A decade later, the 36-year-old corporate vice president, married with a young son and a baby on the way, was surprised to be called up for one-year domestic assignment. The Japanese bombing of Pearl Harbor turned that into four years of wartime service, much of it overseas as a staff officer for General Mark W. Clark, who led the Fifth Army in the invasions of Italy. Ball participated in the Allied landings on Sicily and was air-support officer for the landings at Salerno and Anzio. In *Staff Officer*, a book recounting his military experiences, Ball wrote with fondness: "I was in Omar Bradley's headquarters in Sicily, had dinner with George Patton, tea with the Glaoui of Marrakech, and played golf on his golf courses."

It was not all tea with the Glaoui. Ball and staff officers like him tended to be close to danger and under terrific strain, wrote General Clark, who would later arrange the Korean armistice. "Always hanging over is the ominous knowledge that upon their calculations, their estimates, their plans depend the lives of thousands and perhaps ultimate victory or defeat."[12]

The adventure would end sooner than Ball could guess. In September 1943, in the hills near Salerno, he received a letter from his Uncle George. Its pivotal sentence: "Our feeling is that your services here at home are badly needed."[13]

Ball shared the letter with Clark, who had a sympathetic ear. In the 1930s, Clark had wavered in his military ambitions, taken a leave, and spent several months in Muncie working for the same George Ball, primarily dealing with the wealthy man's various outside interests. Clark ended up marrying the daughter of a Ball salesman. While Clark couldn't furlough Ed Ball outright, he could support his rotation back to the States.[14]

In April 1944 Ball was assigned to Ford's River Rouge plant in Dearborn, Michigan, where he remained until February 1945. He then came home to Muncie and an ailing family business. The fruit-jar market had peaked 15 years earlier, with sales plummeting by more than half as the Great Depression dragged on and housewives reused jars. Many also bought refrigerators: annual U.S. production had leapt from 5,000 units in 1921 to a million in 1931 to nearly 4 million units shortly after the war.[15]

Yet jars were only part of the story. Diversification into beer, liquor bottles and containers for the likes of Gerber and Kraft—not to mention rubber gaskets for refrigerator seals—picked up slack during the 1940s. Ball had also branched into the automotive and aviation markets, for which it supplied a

variety of rubber and metal parts. The company had doubled its total output under the flood of wartime orders—Ball produced more than half the battery casings used by U.S. and allied forces, among other things—stretching both its equipment and workforce. When the tide of orders receded at the war's end, there was no hiding the company's dilapidated state.[16]

Ed Ball was named president of the family business in 1948. Placing a self-effacing, low-key, humble man with a philosophy degree in such a role might seem an odd choice, given the circumstances. "He didn't bowl you over, he didn't overpower you. He wouldn't get decisions made through force,"

Ed Ball

remembered Alexander E. Bracken, Frank Ball's grandson. "But therein lay his strength—such traits describe a character that is just extremely powerful because it's not based on power."

The new leader faced a monumental task. Ball Brothers had developed a reputation among competitors as "the most decadent company in the industry," suffering the public perception that "the company remained one of five elderly gentlemen, probably with long beards, content to rest on their laurels as philanthropists while their obsolete fruit jar business, which had made them wealthy, drifted into oblivion along with buggy whips, celluloid collars and high button shoes," as Ed Ball put it.[17] It would have been easier to start a new company than to rebuild the old one, he felt.[18]

A terrible personal setback would make challenging times even harder for the new corporate leader. In 1949, Ball and his wife Isabel left their three young children in Muncie to join friends in Fort Myers, Florida, where they chartered a fishing boat. The craft was 50 yards offshore when the boat exploded.

"He had this cigarette in his mouth and I said I smelled gasoline," Ball recalled. "That's the last I remember."[19]

The explosion killed his wife and seriously injured Ball. He refused to stay in the hospital, insisting his children have their only parent at home. There, he recuperated for months.

Ed Ball then continued nursing the company back to profitability and led the effort to streamline its glass, metal, rubber and paper operations. He diversified Ball into plastics and, with the help of a Chicago "business finder" named Stuart Cochrane, metal decorating. But Ball had something more in mind.

His long fascination with aviation, his wartime experiences with the power of technology, and the general tide of the times led him to seek modernization and new high-tech markets as vehicles of renewal. He needed brainpower to do it.

Ed Balls' first move was to lure R. Arthur Gaiser away from the Libby Owens Ford Glass Company of Toledo, Ohio, in mid-1955. Gaiser, who had two dozen patents to his name, was to establish a research and development department, which Ball Brothers had done without for 70 years. Rather than build such a technical department from scratch, Ed Ball aimed to buy a company with such capabilities and move it to Muncie—specifically to 90,000 square feet of empty space at the plant.[20] To that end, Ball suggested his new research and development chief pay his business finder a visit.

Cochrane met with Gaiser in Chicago in September 1955. Cochrane explained to his guest that Ball Brothers' expertise appeared to be in molding and forming material, and that Ball would need to establish a well-rounded technical team before the company should dare venturing into areas beyond their historic strengths. Further, he said, acquiring an expanding business with a technical group was "almost an impossibility" given the general shortage of technical personnel. He knew of no shining prospects anyway.[21]

Gaiser considered the man across the desk. "If you don't mind my asking, just how exactly do you go about finding your companies?" he asked.

"Oftentimes," Cochrane answered, "just by hunch."

IN EARLY OCTOBER 1955, Cochrane had a hunch about Ed Safford's weighing device company. Ed Ball, his vice president Bill Schade, and Safford met at Cochrane's Chicago office. So convincing was Safford—a man one contemporary described as "a guy you couldn't help but like"—and so dire Control Cells' financial state, that Ball loaned the Boulder entrepreneur $4,000 of his own money to tide the operation over until the Ball Brothers board met to decide on a formal relationship.

The board gathered on October 25, 1955. George Ball, the company chairman and last survivor of the original Ball brothers, had died three days earlier at age 92.

The directors of the Midwestern glass, zinc, rubber and plastics manufacturer were skeptical about investing in an unknown electronic-scale company in the Colorado boondocks. Anticipating their reluctance, Ball and Schade presented them with something well short of outright purchase. Rather, it would be a complicated agreement boiling down to Ball Brothers spending $35,000 for Control Cells' plant and equipment and an exclusive license to make and sell the weighing devices.[22]

The votes fell six in favor, four against.

Ten days later, Schade sent Ed Ball a note mentioning a Brooklyn engineering firm's estimate that Control Cells would sell $1.7 million in weighing devices in the next six months, including 400 toll road installations and 50 installations at batch manufacturing plants. Control Cells had "very good" prospects of breaking even, Schade said. But Schade also noted "the question I have in my mind about the competence of the organization which allowed the business to reach its present condition."[23]

On December 15, 1955, Ed Ball mailed Safford a $15,000 check. "It is always exciting to enter a new activity and add new business associates, and we at Ball are particularly pleased with this one. We feel that your group fits beautifully with our organization, and together we should have something close to an ideal combination of skills, resources and imagination," Ed Ball wrote in the accompanying letter. "Cordial Christmas greetings from all of us."[24]

For a few months, Safford ran his division with relative autonomy, growing it to about 25 employees and scrambling to keep up with demand. The Muncie parent spent $3,000 on three lots and built a new $60,000 home for its subsidiary at the still-vacant Boulder Industrial Park that spring. But the honeymoon would be brief.

John W. Fisher

In early June 1956, the Ball Brothers operations chief, John Fisher, flew to Boulder. Fisher, a Harvard Business School graduate married to Ed Ball's sister, was an astute and aggressive manager. He didn't like what he saw at Control Cells. Product quality was a particular concern, with failures of "castings, knife edges, spot welds and circuits" causing even existing customers to ask for products on a trial basis. More disconcerting was that the scale's advertised accuracy of about 1 percent of actual weight seemed to diminish with use.

"Safford admitted that he had 'gotten away with murder' insofar as shipping out merchandise manufactured in the cramped quarters which they formerly occupied," Fisher wrote a colleague after the trip. "It might be well to satisfy yourself that Safford fully appreciates the need for a quality product of an instrument of this character. I was not fully satisfied with some of the evasive answers he gave me when I pressed him on this point."[25]

Fisher concluded that a "great deal of realism must be added to this organization before it can be successful."

ON JUNE 28, 1956, Control Cells would unveil itself before its adopted hometown, inviting the entire city to an open house. Ed Ball, the 51-year-old president of the Ball Brothers Company, piloted his twin-engine Piper Apache in for the occasion. He brought with him his children and his lucky mason jar, a companion wherever he flew.

Ed Ball's presence at the Control Cells open house had been well noted. The *Boulder Daily Camera* had published his photo the day before, as well as that of Schade. The printed image of the Ball president, hair Brylcreemed and the soft features of his youth sharpened with age, gave the impression of an executive at the height of his power.

The photos were part of a generous two-page spread under the banner headline "Control Cells to Have Open House Thursday." The official invitation took the form of an advertisement. Ball's and Schade's presence merited its own story, under the headline "Ball Brothers Officials To Attend Open House: President E.F. Ball, Executive Vice President W.C. Schade Visiting Here." Any company coming to town—particularly a technology company—was big news, explained Laurence Paddock, a longtime editor of the town newspaper.

"It was one of the early firms doing something in Boulder," Paddock recalled. "There was a furniture company, Control Cells, and the subscription-fulfillment arm of *Esquire* magazine came in around that time."

Ed Ball personally welcomed many of the roughly 400 locals dropping by Control Cells' new building that evening. But the man from Muncie was more interested in ground truth than public impression. Ball Brothers was pouring $50,000 a month into Safford's operation, with no apparent profits on the horizon.[26] Ed Ball's vision of buying a technical company capable of doing research for the Muncie parent was in peril.

The event was, in fact, the beginning of the end for Ball Brothers' first foray into high technology. In other circumstances, Ed Ball might have sold the new building and written off his Boulder experience as folly with residual tax benefits. But a casual get-together with a University of Colorado physicist would both hasten Control Cells' demise and steer the jar company toward space.

Chapter 4

From Scales to Space

By the time the Control Cells Division of Ball Brothers held its open
house, more than three years had passed since the Rocket Project's Lyman
alpha breakthrough. The University of Colorado group was selling its still-
unmatched pointing controls to Army and Navy solar researchers as fast as
they could make them.

The team building space hardware on Boulder's University Hill had nothing
in common with the Midwestern manufacturer's weighing-device subsidiary
on the outskirts of town. The two efforts would surely have wandered down
their separate paths had not David Stacey, the University of Colorado physicist
and Rocket Project leader, happened to be a friend and neighbor of Ed Safford,
the general manager of the Control Cells Division of Ball Brothers Company.

One sunny afternoon in late June, there was a knock at the door of 601
Baseline Road, the Stacey home. The physicist opened the door to find his
friend Safford standing with a distinguished-looking gentleman. Given the
newspaper accounts of the Control Cells open house, the face might have
struck Stacey as familiar. Safford introduced his patron, Ed Ball.[1]

Stacey invited them to the back porch. The setting was magnificent, with
the Flatirons towering over Boulder's Chautauqua Park just uphill. It was the
domain of someone doing quite well in life.

Indeed, Stacey appeared to have little to complain about. After earning his
PhD in 1954, he had continued working as a University of Colorado research
assistant at a salary higher than that of a new professor. Most new PhDs uproot to find work; Stacey hadn't changed desks. He was 38 years old with a
loving wife and three healthy young children, family money in addition to his
own, and he lived in one of the growing city's finest neighborhoods.[2]

But things were changing, Stacey sensed. William Pietenpol, the CU
physics department chair and titular head of the Rocket Project, had retired
two years before. William Rense, whose Lyman alpha data had brought the
sun-seeking biaxial pointing control to prominence, had taken over the Rocket
Project. Rense found waning support in his own physics department.[3]

Several of Rense's colleagues had quietly decided that the Rocket Project amounted to a manufacturing line taking up space meant for rarefied scientific research. And it was true: the pointing control, although continually refined, had kept its same basic design for years. The skeptics chose to overlook that the Rocket Project, by enabling pathbreaking research, had put the university at the forefront of solar science—not to mention the solar science Rense's group conducted with their own instruments.

Some of the professors also were dubious of the project's reliance on U.S. government money. Washington had yet to become the sugar daddy of American scientific inquiry. The taxpayer-funded National Science Foundation, which spends roughly $7 billion a year on university research today, was a bureaucratic infant with just $30 million to share with the nation's universities.[4] Landing research dollars had been a positive for professors, but until the military and, later, agencies such as the NSF ramped up their spending, the scarcity of research funds had meant few universities could develop their modern addictions to grant money. The Rocket Project was one of the early engines driving what would become relentless pressure on American professors to land research money, establishing "new institutional imperatives for research where none existed before."[5]

The Rocket Project preceded the institutional imperative, at least in the CU physics department. Professors worried about compromised intellectual freedom and the appearance that the military was calling the shots in the halls of academia. The prevailing attitude about the Rocket Project had faded from glowing pride to something close to resentment. With time, despite their contributions, the Rocket Project's leaders had come to feel like second-class citizens.[6]

As Ball, Stacey and Safford sipped their drinks, chatted, and considered the overlapping wedges of Flatirons sandstone, Stacey may well have mentioned such things. The topic would have certainly piqued the interest of Ed Ball.

Ball had first sketched the outlines of a Ball Brothers research group in September the previous year. He wrote his deputies: "Has any thought been given to the feasibility of taking on some government research project as a way to build up our own research organization? I understand that there are government projects available that might help our Research Department become self-supporting much more rapidly than pulling it up by its own bootstraps."[7]

Art Gaiser, the Ball Brothers' R&D director, responded that he doubted the Muncie manufacturer "would be very attractive as a research agency until we have built up our organization to a presentable status, at least on paper."[8]

Control Cells had little kinship with Ed Ball's vision. Something like Stacey's Rocket Project, though, would certainly be "presentable," a fact unlikely to have escaped the visitor from Muncie. That Stacey and Ball had much in common must also have been quickly evident. Both were Midwesterners, veterans, pilots and fathers of three. Both spoke softly and sparingly and preferred to leave the spotlight to others.

Their conversation probably drifted to electronics. Safford had pitched his Muncie boss the idea of his physicist friend offering technical expertise to help fix his troubled company's weighing devices. But what exactly they talked about, no one will ever know.

That summer afternoon could well have been in Ball's thoughts some years later, when he wrote: "The history of the world has turned on events that at times might have seemed unimportant. Luck, fate, circumstances have played their parts. Whims, personalities, temperaments, prejudices of individuals have all helped to weave the intricate pattern that becomes the tapestry of history.

"The same factors shape the destiny of a business enterprise. Far from being a faceless, soulless institution, a corporation is a living thing, the result of many events, the product of so many who have played a part in shaping its destiny."[9]

Ed Ball sent Safford a three-page letter the following week. He thanked Safford for his hospitality in Boulder. Then the hammer fell.

"You have the finest organization and best accumulation of overhead I think I have ever seen without any sales or sales program to support it," Ball wrote. "I was truly shocked to find nothing more tangible in that important facet of a business after the lapse of six months."[10]

Ball continued with an exhaustive request for data on his Control Cells division. He wanted to see a projected breakeven point and salaries "from General Manager down to janitor." He asked for short- and long-range objectives and proposals for achieving them; copies of letters, contracts, drawings and patents; and even "a plan with logical steps and in the proper sequence as you see them of the liquidation and disposal of Control Cells should a reasonable length of time, say at the end of 1956, show that sales cannot be generated in sufficient quantity to justify the continuation of the operation." He urged Safford to "stop chasing will-o-the-wisps, and get down to some hard, basic, fundamental thinking."[11]

Safford responded by stressing how hard his 30-person staff was working and predicting breakeven sales for September. Safford believed deeply in his weighing device and was struggling to make it a success, Ed Ball knew. But Ball was already sketching a new battle plan.

Ball Brothers began pitching contract research services to the likes of Chrysler Missile Operations and the Army Signal Supply Agency, boasting of expertise unrelated to weighing devices or the molding and forming of glass, rubber, plastic and zinc. Instead, the proposals discussed Ball Brothers' sudden competence in such things as "biaxial pointing controls," "servomechanisms," "spectroscopy aspects of biaxial pointing controls," and "ultraviolet and visible spectroscopy." David Stacey was listed as a consultant. Stacey even flew to Philadelphia with a Ball group in October 1956, to meet with General Electric's Special Defense Products Department. General Electric, which had won a portion of the Air Force's intercontinental ballistic missile program, asked Ball Brothers for several bids, including a detector to sense micrometeorite strikes and studies of materials that could shield a warhead from the heat of atmospheric reentry.

Ed Ball flew to Boulder in early November to make his move. Stacey and his colleague James Jackson agreed to leave their university positions and join Ball Brothers. Ball would build his second Boulder venture from the ground up, with help from the cash-cow pointing control he intended to make his own.

Jackson and Stacey couldn't do it alone. At the top of their most-wanted list was Pete Bartoe, the former graduate student who had helped design the CU pointing control.

Bartoe had since moved to California, where he was working for the Santa Barbara Research Corporation. There, among other things, he had built upon his pointing control work to design an automatic sextant—a play on the navigation tool used by sailors for centuries—to spot the sun and orient a camera aboard a high-altitude balloon.

Stacey, calling from his home in Boulder, offered Bartoe a job. Bartoe was interested, but said he would need a month's vacation—a lavish amount. He needed two weeks to hunt, a week to visit family, and a week to just do nothing, he reasoned. Besides, it had been the norm on the CU Rocket Project, as Stacey well knew.

Stacey asked Bartoe to hang on and, as Bartoe recalled, repeated his request to someone else apparently in the room. "If that's what he thinks he needs," the voice said. Ed Ball, in the Stacey kitchen, had sealed it. For decades, Ball's Boulder engineers got four weeks of vacation. This "unimaginably generous" perk, as one engineer described it, swayed many career decisions.[12]

On November 6, 1956, the day Dwight D. Eisenhower was reelected, Jackson met with University of Colorado President Ward Darley to announce his and Stacey's departure. Darley, who had become accustomed to watching Air Force checks land in the university's bank account, did not share the physics department's disaffection for biaxial pointing controls. Darley wrote a three-page letter excoriating Ed Ball for his piracy of CU talent. "If the coming of such research agencies as yours results in stripping one of our important operations of its technical personnel, it would be to your immediate advantage," Darley wrote. "But, in the long run, activities which would result in weakening the scientific and research departments…[weaken] us as an important training center for those young men upon whom industry depends for its engineers and research staff."[13]

Ball's reply was conciliatory but firm. He explained Ball Brothers' struggle to lure R&D staff to Muncie and how Boulder's attractiveness to technical talent was a deciding factor in his company's investment in the community. He reminded Darley that Jackson and Stacey were planning on leaving the university anyway, and that "we should be remiss in obligations to our stockholders if we did not offer them a position…."[14]

Bartoe helped Stacey carry file cabinets full of pointing-control design documentation from the building. They loaded them into the Stacey vehicle and drove them to the new Control Cells building, where Safford's crew had made room for them.

"Jesus, Dave, are we doing something illegal, here?" Bartoe asked his boss.

"No, not a problem, don't worry about it," answered Stacey, straining a bit under the weight of all that paper.[15] There had been no contract, no technology licensing agreement. It was tech transfer by station wagon, intellectual-property acquisition by pillage.

On Monday, December 3, 1956, Ball Brothers Research Corporation was formally established, with Jackson as director and David Stacey technical director.

The Certificate of Incorporation listed manufacturing first and research second among the new organization's aims. This seemingly innocuous subtlety, typed on a legal document and quietly filed away, in fact hinted at Ed Ball's ambitious vision for his latest endeavor in Boulder.

Ball had once underlined a sentence in a *Business Week* article. It read: "[C]ompanies that plow back 3% to 4% of their sales into R&D tend to have yearly growth of 10%."[16] Yet neither he nor the Ball Brothers board had the stomach to spend the requisite $2 million a year on R&D, an amount roughly equal to his company's annual net income. Ball wanted the 10 percent growth without the 4 percent cost. He was trying to squeeze water from a stone—an ambiguously defined stone at that.

Then, as now, there were two main types of research and development organizations: the in-house cost center and the independent research house. AT&T's famed Bell Laboratories was the foremost example of a cost center. Narrowly viewed, it was a financial drain. But its discoveries paid off handsomely over time as Ma Bell commercialized or licensed innovations such as the transistor, the laser, solar cells, the touch-tone phone and the fax machine.

An independent research house, by contrast, had no corporate parent. It either sold research services to other companies or fished for government research grants. Modern examples include such giants as the Southwest Research Institute, which, for a fee, studies everything from automotive technologies to space science; and the Battelle Memorial Institute, which in addition to conducting diverse industrial research, runs several U.S. Department of Energy laboratories. Both happen to be nonprofit organizations.

Ball Brothers Research was to be a miniature Bell Labs–Battelle hybrid. Ball would hire glass and other materials experts in Muncie to staff a cost center aimed at improving Ball's products and processes. At the same time, it would have its rocket-centric group in Boulder turn profits from government and commercial work—in addition to random inside jobs for the Muncie parent.

Ball and his R&D chief Gaiser intended to "assemble a nucleus of experts in a few scientific fields, supply them with some instrumentation peculiar to their fields, and thereby attract sponsors for contractual research."[17] The challenge of tantalizing corporate sponsors with such scant offerings seems not to have entered their minds.

Aerial view of the Ball Brothers Research building in 1957

Such was the company Bartoe joined in late 1956. Working from a rented home where shipping crates were a dominant furnishing, he thought through various General Electric proposals.

Others from the Rocket Project would follow, among them Russ Nidey, the photoelectric-eye expert who was, by early 1957, working on his PhD while managing biaxial pointing control work at the University of Colorado. When Ball came calling, Nidey asked one of his professors for his opinion.

"You'll be making 30 percent more than I'm making," the advisor answered. "What do you think you ought to do?"[18] Nidey left his studies to work on pointing controls at Ball.

Bartoe was also asked to evaluate the Control Cells weighing device. The fruits of his work, co-authored by Jackson and Stacey, were published in a brown folder. "A Brief Technical Survey of the Control Cells Systems," dated January 18, 1957, painted a dire picture of the company Ed Ball bought. The three concluded that "the product has been marketed while in the development state, without sufficient laboratory tests, field tests, production facilities and quality control...in a scramble-gamble to get a marketable product that would sell like hot cakes."

They found customer files full of rework and repair, a calibration system that did not calibrate, and an alarming tendency to introduce new products while the original scales remained defective. "One is reminded of the horse in a very lush pasture that leans on the fence to eat the grass on the other side," Bartoe and colleagues wrote.[19]

Control Cells had violated what Ball Brothers Research's founding engineers considered a fundamental principle of engineering design: "namely, make one [product] as best you possibly can with precision components throughout—then, when the bugs are removed and the operational characteristics thoroughly understood—redesign for production."[20]

On February 2, 1957, the Boulder newspaper published a short article titled "Research Project Halts Production at Control Cells." Control Cells, it read, "announced today that important improvement possibilities in its basic product have been uncovered. As a result of the discoveries all marketing and production efforts of the present product are to be suspended immediately. The Ball Brothers Research Corporation, which jointly occupies the Ball Brothers building with the Control Cells Division, will perform the development work..."

Ball Brothers Research hired several Control Cells men. Ed Safford was not among them.

"This is just a note to express my sincere regrets in seeing our joint efforts concerning CONTROL CELLS come to an unsuccessful conclusion," Safford wrote to Ball. "The very great problems in pioneering a radically new product were, perhaps, more severe than either of us fully appreciated."[21] It was an insight that would hold true in this corner of Boulder for years to come.

CONTROL CELLS HAD BEEN A DISASTROUS INVESTMENT for the Muncie jar company. But rather than fold and quietly return to his container business, Ed Ball had doubled down. Ball Brothers Research grew to roughly 30 employees by mid-1957. The new engineers and technicians had been hired mainly to build pointing controls. In addition to file cabinets, Stacey and Jackson had brought with them a contract with Air Force Cambridge, which had been the CU Rocket Project's first customer. Although still a CU customer, Air Force scientists needed more hardware than the university's Upper Air Laboratory could build. Ball Brothers would simply expand the supply of the $25,000 pointing controls, which remained hand-crafted objects that took months to build and test. Ed Ball had the government work he had hoped for.

By early 1958, Aerobee rockets carrying Ball Brothers Research pointing controls built from University of Colorado designs were leaping above the stratosphere like fish over a lake. A company famous for its mason jars was at the edge of space, but struggled to bring in enough business to justify its existence as a for-profit research enterprise.

Although the General Electric proposals hadn't panned out, the Boulder startup landed work elsewhere, designing gamma-ray detectors and Geiger counters for uranium prospecting and selling them to the Turkish, Iranian and Yugoslavian governments.[22] For the Muncie parent, they created a plastic bottle-blowing machine, Ball's first foray into plastic containers. They developed a means to link the guidance system of a Navy Regulus cruise missile to that of the B-52 bomber carrying it, an effort cut short by the Navy's cancellation of the Regulus program. For Air Force Cambridge, Boulder engineers made a special pointing control for use on a high-altitude balloon.[23]

Jackson had also launched an eclectic research program, including investigations into anesthetic monitors and ramjet engines. Art Gaiser, the Ball Brothers R&D director, killed them, as well as the effort to salvage Safford's control cell. Concerned by the absence of the industrial research program he had envisioned, Gaiser warned Jackson of banking "almost entirely on the military to support us."[24]

Ball Brothers Research had lost $40,000 on $264,000 in sales in 1957. Three-quarters of the business came from pointing controls and other military projects.[25] Ed Ball felt pressure from his board. He explained to his Muncie directors:

> This is the newest member of our industrial family, and it is difficult to say what its future may be or what specific benefits it may bring to the company as a whole. It is our feeling, however, that any progressive industrial organization today must engage in a substantial research program to improve processes and develop new products....
>
> Through a shortsighted policy, we might be able to cut down considerably on our overhead and improve current profits by the reduction of our research program. This course, however, I believe would be a serious mistake, and particularly so since it is now so close to operating on a break-even basis.[26]

Given his complicated mandate from Muncie, Jackson would have struggled to win Gaiser's favor had he executed flawlessly, which he had not. Jackson's tendency to promise customers more than the company could deliver also darkened Gaiser's view of his Boulder chief. Gaiser fired Jackson in February 1958, banishing an American space pioneer from the sun-seeker program he had been instrumental in developing. Jackson moved to New Mexico and spent the rest of his career with the Army Missile Test and Evaluation Command at White Sands Missile Range.

With Jackson gone, Gaiser assumed Stacey would take over the top job. But Stacey declined, preferring to remain technical director. Ruel C. "Merc" Mercure succeeded Jackson, with clear instructions to make Ball Brothers Research profitable.[27] Mercure was 25 years old and a lanky six-foot-five inches tall. He "looked like a confused teenager," as a colleague later put it.[28]

Ruel C. "Merc" Mercure

Looks deceived. Mercure, born in the town of Loveland northeast of Boulder, had earned his physics PhD under Rense in 1957. He had designed and flown, on a CU pointing control, the first device ever to photograph the sun in the ultraviolet. His instrument had been a descendant of the monochromatic camera Rense had abandoned late in the hunt for Lyman alpha five years earlier, and was a shining example of the interplay between advancing technology and scientific progress. Better tools enabled new experiments; new experiments yielded new discoveries; new discoveries begged new questions whose answers demanded better tools yet. A perpetual cycle of scientific inquiry, which had played out on the ground for centuries, was now a fundamental driver of space technology.

Prior to returning to CU for graduate school, Mercure had worked for the new Rocky Flats nuclear-weapons plant south of Boulder as well as for a secret government project attempting to make an atomic bomb based on ultracold liquid hydrogen. Mercure had even traveled to the South Pacific for a competitive midnight test against the "dry" hydrogen bomb being developed at Los Alamos.

The dry bomb went first. The granddaddy of the world's nuclear arsenal lit up the sky "just like it was noon," Mercure remembered. Liquid hydrogen would produce a flickering candle by comparison, the team knew. They packed up their bomb and went home, never to test it.

Mercure had first encountered Ed Ball a year earlier, in the east Boulder structure built for Control Cells Corp. "I was in the machine shop, and here comes this guy wandering through the back door. And I said, 'I'm sorry, you can't come in here, you'll have to go around to the front.' And the guy said, 'Oh, I'm sorry,' and he went around to the front and introduced himself," Mercure recalled.

"Somebody ran up to me and said, 'Mr. Ball's out there,' and I came around and there was the guy I just threw out the back door."[29]

The research group Mercure now led bore little resemblance to the company that would later take aim for a comet. The transformation of the jar company's strange side business into a top American space contractor would be no less improbable than the Ball Brothers Research creation story itself. John Lindsay, far removed from Boulder and Ball, would trigger the metamorphosis.

Chapter 5

Prelude to a Satellite

S earch the Internet for John C. Lindsay, and the name of an award pops up: The John C. Lindsay Memorial Award for Space Science. It recognizes the NASA Goddard Space Flight Center employee "who best exhibits the qualities of broad scientific accomplishments in the area of Space Science." John Lindsay's most apparent legacy is thus his name as presented in other scientists' curricula vitae.

In a way, it's a fitting tribute to someone who, as one writer put it, "was powerfully motivated by an admirable but tormenting ambition to win recognition in science and leave his mark on history."[1] Lindsay, whose accomplishments have faded into obscurity, should be counted among the visionaries of the American space program.

Lindsay smoked a pipe and lived on black coffee. He was big, quiet, likable, and worked independently toward audacious goals. A colleague described Lindsay as "a Scot. He's honest and speaks his mind. He's a canny in-fighter, a courthouse-steps politician who knows how to get things done, which is unusual for a scientist. He reacts very hard to the smallest opposition.... John never lets his right hand know what his left hand is doing. This is how he's able to ride off opposition."[2]

A native Virginian, Lindsay spoke slowly and asked penetrating questions. They often related to Lindsay's job and passion: scientific hardware used to study the sun.

Lindsay first studied the sun with the Naval Research Laboratory, where he worked until NASA opened its doors on October 1, 1958. Then he became one of the first space scientists at NASA's new Goddard Space Flight Center in Greenbelt, Maryland, which was born with its parent agency.

As with many solar scientists, Lindsay's scientific work depended on views of the sun from above Earth's stratospheric filter. He had worked on Naval Research Laboratory instruments that rode sounding rockets, using their data to try to understand the showers of high-altitude X-rays and ultraviolet radiation coming from solar flares. In fact, the Navy lab was the first to identify X-rays—and not the Lyman alpha ultraviolet the University of Colorado team

John Lindsay, lower right, with fellow Naval Research Laboratory scientists as they analyze solar data

had briefly been famous for measuring—as the ingredient in solar flares that causes radio static.

Lindsay was recognized at NASA as both an excellent scientist and an effective project manager. It was a rare combination, said Nancy Grace Roman, the NASA scientist charged with ensuring that early NASA science missions delivered the data they said they would.

"He came from physics rather than astronomy," Roman said. "Physicists end up doing experiments and astronomers can't. He understood instrumentation, he had done rocket work, and so he knew how to do these things. And at the same time he was well-enough respected both in the scientific community and within Goddard that he had a fair amount of power simply from that."[3]

Lindsay was familiar with Boulder-built biaxial pointing controls. He knew David Stacey and others on the University of Colorado Rocket Project, and he knew Ball Brothers Research had gotten into the business of tracking the sun.

From day one at NASA, Lindsay had his sights set on launching a sun-seeking satellite. He set about his quest at the perfect moment. In the late 1950s, despite years of sounding-rocket shots, the sun remained poorly understood. As had been the case a decade earlier when the University of Colorado Rocket Project had begun, scientists had a particular interest in the sun's upper atmosphere—the corona. They believed it held clues about the mysterious magnetic powers that ultimately sent blasts of high-energy radiation toward Earth.

V-2 rockets had first parted the atmospheric curtain for science in the late 1940s, and the CU Rocket Project provided the first steady view of the sun with its Lyman alpha shot in late 1952. But sounding rockets were only good for a few dozen seconds of sun spotting at a time. Solar researchers still felt like zoo visitors pausing at the chimpanzee enclosure. They wanted to be more like

Jane Goodall, observing the objects of their scientific desire for weeks, months, even years on end.

To maximize the value of the technology they had, scientists got creative. James Van Allen, who had left the Applied Physics Laboratory for the University of Iowa, came up with a combination of a high-altitude balloon and a sounding rocket he dubbed the Rockoon.

In 1956, a Naval Research Lab team led a 650-sailor expedition off the California coast, complete with a 450-foot dock landing ship and accompanying destroyer. In the dock landing ship's bay were three truckloads of helium and parts for 10 Rockoons.

The setup was complicated. A balloon floated to 80,000 feet and waited with its rocket dangling below from a nylon line. A radio truck parked on deck tapped into broadcasts from Tokyo, Mexico City, San Francisco, and New Mexico. If all four radio connections faded at once, the sun's interference with the ionosphere had to be the culprit, scientists figured. Navy scientists pushed a button and the rocket ignited and shot another 60 miles skyward, looking for X-ray and ultraviolet solar flare signatures all the way.[4]

It was a rather involved way to gather scientific data. Plus, Rockoons and other sounding rocket efforts tended to raise as many questions as they answered. Years later, a Ball engineer named Dick Woolley would explain the state of affairs in musical terms:

> Suppose that we longed to understand all about an orchestra: the theories of harmony and counterpoint, the structure of symphonies, even the details of construction of oboes and violins. But suppose that we could learn about all this only by listening to the sound of an orchestra through a wall. And suppose the wall blocked out most of the sound, so that only middle C came through clearly, C-sharp and B came through badly muffled, and the rest of the notes not at all. With years of careful study we could frame a few uncertain notions about the nature of an orchestra.
>
> Now suppose that a door in the wall opened for an instant, then closed again. In that instant of clear hearing, we would gain more real information than in all our years of straining our ears at the wall; but we still wouldn't know what the information meant. We'd probably feel more ignorant than ever, because the new facts would wreck our old theories.
>
> If the door opened again for another instant, we would be surprised: The orchestra would sound very different now. For the first time, perhaps, we would suspect that change is the essence of music, and we would know that we must somehow block open that door long enough to hear the whole symphony.[5]

The best way to prop the door open, John Lindsay figured, would be with an orbiting observatory to study the sun.

Political developments surrounding the dawn of the space age happily aligned with Lindsay's aims. In 1946, the RAND Corporation had been directed to study the possibility of American orbiters, and by the mid-1950s U.S. leaders had committed to launching a satellite.[6] The need was much more strategic than scientific.

In addition to a rocket-mounted nuclear arsenal to deter the Soviet missile threat, the Eisenhower administration desperately wanted reconnaissance satellites capable of watching Soviet nuclear-weapons sites. The hope was that eyes high over the Iron Curtain would lessen what Eisenhower viewed as an unacceptably high risk of America suffering a Soviet first strike. Yet despite overwhelming national-security concerns, the White House chose to open the door first to space science instruments.

It was a decision of statesmanship. Earth orbiters would push legal as well as technological frontiers. Eisenhower bristled at flouting international law (despite American high-altitude-balloon and U2-spy-plane flights over Soviet territory routinely doing so). Rather than launch a spy satellite and risk geopolitical uproar, Eisenhower figured it was safest to establish precedent with benign science missions and then, with "freedom of space" established, start space-based spying in earnest.[7] Further supporting such an approach was the upcoming International Geophysical Year (IGY). Lasting from July 1957 to December 1958, the IGY aimed, among other things, at better understanding the sun-Earth relationship during a peak of the 11-year sunspot cycle. For the occasion, the U.S. Army, Navy and Air Force had all proposed science satellite programs—Explorer, Vanguard and Pioneer, respectively.

The Soviets would blaze Eisenhower's legal trail for him. On October 4, 1957, a two-foot silver sphere, its antennas sticking out like a seal's whiskers, rose from the desert steppes of Kazakhstan. Its harmless beeping became a clarion call to Cold War action, triggering a national uproar about the state of American science and technology. The space race was on. The gears that would ultimately produce NASA began to turn, and although its manned program got the most attention, the agency enjoyed a billion-dollar budget for science satellites when it launched in 1958.[8]

Within a year of Sputnik I, America had retaken a lead in space science it would never relinquish. Following a disastrous rejoinder to Sputnik—the prototype Vanguard I exploded on the launch pad in early December 1957, a nationally broadcast fiasco the press would label "Kaputnik"—the U.S. Army's Explorer 1 reached space on January 31, 1958.

Wernher von Braun's second-generation V-2, dubbed the Jupiter C, carried it Explorer 1, which in turn carried a Geiger counter built by James Van Allen, solar scientist and Rockoon inventor. The radiation-detecting instrument crackled amid what would be named the Van Allen radiation belts. The American space science enterprise basked in the discovery's glow. U.S. reconnaissance satellites soon followed, but the political and scientific importance of research satellites had been well established by the time Lindsay formulated his plans for a satellite to study the sun.

Lindsay's idea for an orbiting sun-follower found support in the scientific world. Solar research was high on the lists of the groups formulating the nation's emerging space science agenda. In its first science satellite recommendations in 1958, the National Academy of Science's Space Science Board suggested developing a space-based solar observatory.[9]

The sun also happened to be an easy target. The brightest object in the sky was hard to confuse with another star. Modern space telescopes such as Hubble nimbly steer from galaxy to star to nebula, heeding scientists' requests with graceful precision. In the late 1950s, the technologies behind such agility didn't exist.

Then there was the human factor. Lindsay had been one of about 50 Naval Research Laboratory scientists who switched allegiance to NASA's new Goddard Space Flight Center in 1958. The Navy lab had been a leader in sounding rocket solar research since the V-2 days, and the erstwhile Navy scientists formed the nucleus of NASA's space-science organization.[10] Indeed, NASA's first space science chief was Homer Newell, the Naval Research Lab scientist who had met with William Pietenpol and his CU Rocket Project leaders at White Sands in early 1948, days after Pietenpol had agreed to build a coronagraph on a rocket. Newell, no stranger to CU's—and later Ball Brothers'—work supporting solar science, harbored ambitions for a solar observatory even grander than Lindsay's.[11]

So the stage was set. Directly aiding Lindsay's cause was Goddard chief Harry Goett's tendency to run something of a loose ship. The scientists in the new NASA center "approached their work in a highly individualistic manner. They questioned everything, including orders from above.... The accomplishment of an experiment that produced significant new information was what counted; costs and schedules were secondary," Newell said, adding that "the seemingly casual approach of the Goddard scientists looked too undisciplined to work."[12]

At a golden moment scientifically as well as politically, Lindsay also enjoyed what modern NASA scientists would consider unheard-of license to do as he pleased. He could pick his own multimillion-dollar mission and select the contractor of his fancy from his Goddard office. NASA wouldn't get around to reining in such behavior for another two years, when it consolidated the power to select missions at NASA headquarters, where it would remain.[13]

JOHN LINDSAY MAY OR MAY NOT have explicitly understood all this, busy as he was managing the Explorer 6 satellite program. An opportunist on a mission for a mission, he was interested in politics only to the extent it served his science. He envisioned a spacecraft with technologies and instruments similar to those on sounding rockets.[14] The prospective solar satellite's ability to point accurately at the sun would be paramount. He turned to industry for ideas on how an orbiter might manage the problem of pointing in space.

He met with the Perkin-Elmer Company, which suggested using internal gyroscopes. Lindsay dismissed them as too expensive and technically ambitious. An aircraft armament company advised using a "bullet belt or bandolier of cartridges" around the spacecraft's midsection, one of which would fire on occasion to keep the satellite's aim true.[15] A third suggested a combination of photoelectric sensors and servomechanisms.

Lindsay flew to Colorado on January 12, 1959. There, in a small conference room in a building east of Boulder, he sat down with David Stacey, Merc Mercure, and others from Ball Brothers Research. Lindsay asked Ball to prepare five proposals, ranging from a system to stabilize Aerobee rockets to the object of Lindsay's affection, which he called a "satellite astrostat."[16] This satellite astrostat, he explained, would be a pointing control.

The men nodded.

But this pointing control, Lindsay clarified, would ride into orbit, bolted onto the third stage of a big Thor Delta rocket. The instrument this satellite astrostat carried would operate above the ozone layer for days or weeks straight. Could they build such a thing?

The men nodded again. But they had no idea if they could build such a thing.

They had to take the risk. The leaders of Ball Brothers Research saw this "astrostat" as a way to keep the lights on.

Stacey asked Pete Bartoe, then 31, to write a proposal, with help from a physicist named Fred Dolder. Dolder had been Bartoe's freshman-year roommate at the University of Colorado. Now, the two shared a small, spare office with gray metal desks, gray metal file cabinets, and a chalkboard.

Fred Dolder

Pete Bartoe

Bartoe worked on the system design; Dolder handled the electronics. Bartoe approached his work systematically. He explored dozens of solutions. He would get comfortable with a design, he said, only after new attempts circled back toward earlier ones. With the orbiting pointing control, the circles never seemed to close. Unlike an Aerobee rocket, whose behavior was undisciplined but predictable, an orbiting, burned-out rocket stage would be just plain erratic.[17] For a couple of weeks, Bartoe wrestled with the problem of a pointing control flexible enough to target the sun aboard a tumbling platform. An instrument bound to the rocket stage would require a pointing arm with the flexibility of an elephant's trunk. There was just no way.

But luck favors the prepared. In the weeks before Lindsay's visit, there had been a lull in business at Ball Brothers Research. Bartoe immersed himself in a mathematical study of gyroscopes. When a gyroscope is a gyroscope, there is stability. Bartoe wanted to know when a gyroscope is *not* a gyroscope.

Wheels on a bicycle at rest are not gyroscopes. Bike wheels' gyroscopic qualities emerge with speed. There is a moment at which—magically, it seems—one can suddenly ride no-handed. Bartoe wanted to know when that was. He was interested in gyroscopes for their potential to provide stability in space. Curious about the mathematical relationship between wheel diameter, wheel mass, and gyroscopic strength, Bartoe hunted down a book at the University of Colorado library, engineering pioneer Stephen Timoshenko's 1948 *Advanced Dynamics*. He pored over Timoshenko's treatment of gyroscopic behavior into the night.

So in early 1959, as he struggled with Lindsay's orbiting pointing control, Bartoe had gyroscopes spinning in his brain. They steered him to an ingenious solution. Bartoe walked into the office one morning and announced to his old friend Dolder: "By God, I've got it solved!" He picked up a piece of chalk and scratched out an image that would appear, more or less, on Ball's proposal for a "Satellite-Borne Solar Pointing Control."[18] Gone was the rocket stage. In its place was a spinning platform—he called it a "wheel"—to help the satellite generate its own inertia.

Configuration-Satellite-Borne Astrostat

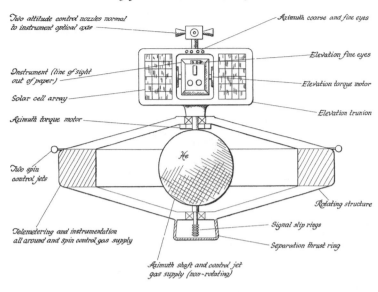

Two attitude control nozzles normal to instrument optical axis

Azimuth coarse and fine eyes

Elevation fine eyes

Instrument (line of sight out of paper)

Solar cell array

Elevation torque motor

Azimuth torque motor

Elevation trunion

He

Two spin control jets

Rotating structure

Telemetering and instrumentation all around and spin control gas supply

Signal slip rings

Separation thrust ring

Azimuth shaft and control jet gas supply (non-rotating)

Visionaries in any field, be it politics or art, see the world not as it is, but rather as it should be, creating new realities obvious to the rest only in hindsight. The same holds for great engineers. The engineer, though, has it harder yet. Before introducing his work to the three-dimensional world, his vision must run a gauntlet of equations proven over centuries to reasonably represent the universe we live in. Intuitive genius alone falls short. Bartoe worked the numbers, producing an elegant design that satisfied them. If the wheel weighed enough and revolved once every two seconds or so, the gyroscope would be a gyroscope, Bartoe calculated.

The satellite's wheel would be ballasted with control electronics, radio equipment, and batteries. The wheel would stabilize the craft along its main axis. Helium-gas jets would stoke the wheel in case friction slowed its motion. Similar to a Rocket Project pointing control, photoelectric "eyes" would spot the sun and tell servomotors which way to point the top half of the craft (which Bartoe called the "sail") to aim a single instrument.

The sail would sport a one-square-foot patch of solar cells supplying a few watts of energy to two servomechanisms, each of which could run on less power than a nightlight. One servo would spin the wheel; the second would aim the instrument. Unlike a biaxial pointing control, the satellite could take its time finding the sun.

Bartoe designed a pair of jets on top of the sail that would puff out helium, nudging the nose of spacecraft up or down until it faced the sun. The jets also would adjust for changes in the sun's relative position during the satellite's time in orbit as well as for the satellite's precession, or the tipsy yawing of a spinning body.[19]

The gyroscope's spin steadied one axis and the gas jets tweaked the second. The remaining axis, called the roll axis, could be left alone. The satellite's instrument could watch the sun regardless of its roll-axis position, much as someone can watch TV sitting up or lying down.

As a final touch, the instrument itself would be aimed by an elevation arm steered by its own "fine eyes," just as on the CU pointing controls. Combining the stability of the gyroscopic wheel, the gas jets, and the elevation arm, the spacecraft could orient its instrument with an accuracy of a minute of arc, Bartoe estimated. That was sixty times better than the one-degree accuracy of the steadiest satellite ever to fly.[20]

Bartoe called Lindsay and talked through the idea. It was a major departure from Lindsay's initial notion of a pointing control on a rocket stage, and with unclear consequences.

"He jumped on it," Bartoe remembered. "He said, 'That's a great idea. Let's do that.'"

On February 20, 1959, Bartoe and Dolder sent Lindsay their proposal. It was 27 pages long, plus appendices. Appendices included mathematical proofs, detailed resumes of 18 of the company's roughly 40 employees, and a nine-page list of virtually every piece of equipment in the 8,000-square-foot building Ball Brothers Research had inherited from Control Cells Corp., down to the power hacksaw and "Assorted Lenses, Prisms and Filters."

They would build and test three satellites in a year, the Boulder group promised. While reflecting their optimism, the proposal showed how little they understood about the complexities of flying a pointing control on the back of a giant, orbiting gyroscope. Bartoe, Dolder and the rest at Ball Brothers Research were going to learn this lesson the hard way.

Chapter 6

Engineering for the Unknown

On a September evening in 1959, Fred Dolder and Merc Mercure flew back to the Rockies from Washington, D.C., with a $250,000 deal. The money was to be spent within a few months on a preliminary design of the spacecraft Pete Bartoe had dreamt up. Another $600,000 or so was promised for three Orbiting Solar Observatories.

It was a windfall for Ball Brothers Research, which had grossed $758,000 the prior year.[1]

"All the way back to Denver, Dolder and I were trying to figure out how in the shit could we spend eight-hundred-and-some-odd thousand dollars," Mercure said. "That was such a large amount of money, we had no idea."[2]

Tired of Ball Brothers Research Corp. scarcely breaking even on its work, Bartoe had included plenty of padding in the proposed budget, to the point that "[w]e thought that was about the biggest number we'd ever heard of."[3]

But a decade earlier, the University of Colorado's Rocket Project team had learned how costly attempting to build something new can get. The experience of Control Cells in this very building had taught a similar lesson. Ball Brothers Research's Orbiting Solar Observatory program was soon vastly over budget, as well.

It wasn't entirely their fault. In March 1959, Lindsay had submitted his initial proposal to NASA headquarters. Two months later, Lindsay had assembled two dozen of the nation's top solar scientists from universities, the military, and NASA for feedback on his satellites.

Each observatory would weigh about 300 pounds and orbit 300 to 500 miles up. Half that weight could be dedicated to a single experiment, he told the scientists gathered in the nation's capitol, and the satellite should have a minimum lifespan of one month.

Lindsay had considered one month a good, long time for the satellite to soak in the sun, compared to the glimpses afforded by sounding-rockets. The scientists, though, thought that if a month was good, six months would be better. The group agreed that "the lifetime problem should be investigated very thoroughly."[4]

Dolder, who had been appointed Ball's Orbiting Solar Observatory project manager, sat in Lindsay's Goddard Space Flight Center office. Dolder had a background in solar research, having worked for Walter Orr Roberts at the High Altitude Observatory in Climax, Colorado, for three years prior to entering the aerospace business. For his physics master's thesis, which Roberts supervised, Dolder had pored over hundreds of glass-plate images to draw a link between solar flares and a particular line in the sun's spectrum.[5]

"You know, if you can make this thing last three weeks, why can't you make it last six months?" Lindsay asked.

"Why can't we make it last six months?" Dolder repeated, as if doing so might make the idea go away.

Lindsay, who had just asked for unprecedented longevity in the vacuum of space, exhaled cigarette smoke into the awkward silence.

"Well, we could try," Dolder finally said.[6]

The Ball project manager later recalled that "[t]here was really no formal documentation or really anything. We just sort of changed the rules to make something that would last."[7]

Conventional lubricants such as oil and axle grease evaporated in a vacuum, leaving metal-on-metal behind. Something called cold welding can ensue, where the surfaces in contact fused together as if under an acetylene torch. Bartoe and Dolder had recognized the problem while crafting the proposal for Lindsay. No satellite had ever flown with even a single moving part exposed to space.

They had decided to seal gears and axles from the vacuum. It would work for about three weeks, they guessed, before the lubricant leaked and the satellite's rotating wheel ground to a halt. None of this was mentioned in the proposal. But for that reason, they had pitched Lindsay a spacecraft that would survive just three weeks.[8]

The satellite Lindsay had dangled before the scientists also dwarfed what Ball had proposed. Bartoe had estimated the satellite would weigh 150 pounds, 30 pounds of it being the instrument. But a Thor Delta rocket, its planned ride, could lift 300 pounds to low-earth orbit. So Lindsay decided to add another instrument.

But adding instrument weight to the observatory's sun-facing sail forced more mass into the rotating wheel to keep the gyroscope a gyroscope. No problem, Lindsay figured: he could add spin-tolerant instruments to the wheel, too.

More instruments needed more power, requiring larger batteries and more solar cells on the stationary sail, which led to still more weight. In all, the satellite's load expanded from a single spectrograph to 13 instruments, by far the most ever to be carried into space on a single satellite. And so it went, Dolder recalled, often without a memo changing hands.[9]

"The problems were quadrupled, but we were too dumb to know it," he said.[10]

Two instruments were to be pointed at the sun. One was Lindsay's, designed to measure a set of wavelengths then classified as soft X-rays, now called extreme ultraviolet. The other was to be developed by a University of Colorado team led by William Rense. Building upon scientific successes beginning with the first successful Rocket Project pointing control, Rense aimed to learn more about Lyman alpha radiation.

The rotating wheel was to carry photodiodes and ion chambers designed to capture different ultraviolet wavelengths, gamma ray detectors, a neutron monitor, a proton-electron detector to better understand the Van Allen radiation belt, and a NASA experiment testing the behavior of various metals and paints in space.

Lindsay's original notion of a simple, orbiting pointing control had morphed into a creature worthy of its ambitious name. This would be the world's first observatory satellite, and it would pioneer the use of instruments as interchangeable cargo. NASA called the Orbiting Solar Observatory "the first of the 'streetcar' satellites because it has a series of experiment apparatuses aboard as 'passengers'—thirteen of them."[11] Such streetcars are now known as satellite buses, which in their various forms remain the mainstay vehicles carrying scientific instruments into space.

There were other technical challenges. The telemetry system—the satellite's means of collecting, handling, and sending instrument and spacecraft data to the world below, as well as receiving instructions from mission control—was mistakenly assumed to be NASA's job and thus absent in the proposal. In addition, the satellite would have to store data for later transmission, something never before done in space.

There was also the problem of temperature. The streetcar satellite would, in each 90-minute orbit, have to survive the equivalent of an Arctic winter and a Death Valley summer without onboard heaters or air conditioners.

When they had sent in their bid, the Ball Brothers Research group had no idea they would have to solve such problems—and many others—to fulfill their promises to John Lindsay.

"It's very nice to be ignorant. We never figured out that there was any great problem in doing it," Mercure recalled. "Well, there *was* a great problem in doing it."[12]

•

BALL NEEDED ENGINEERS to work on such problems. The company went on a hiring binge that started before the Orbiting Solar Observatory contract was signed. Among those brought in was Dick Woolley, who as a University of Colorado student had once watched Rocket Project Professor William Pietenpol practice his no-look chalkboard circles.

Woolley had worked for North American Aviation's Rocketdyne division and then the Glenn L. Martin Co. (later part of Lockheed Martin), which in 1955 had launched a major operation in Denver. Martin hired hundreds of workers in Colorado to design and build Titan intercontinental ballistic missiles for the Air Force.

Martin had made the leap from biplane startup to prime military contractor in 1914, when the U.S. Army Signal Corps ordered 14 propeller-driven planes from a company named after its flamboyant, barnstorming founder.[13] The company had learned mass production under the military's wing, building nearly 9,000 planes for the Army and Navy during World War II, more than ten times Martin's total output in the two decades before the war.[14]

The boom ended abruptly. U.S. aircraft sales fell from $16 billion in 1944 to $1.2 billion just three years later.[15] Viewing missiles as the future of military aviation, aircraft companies aggressively sought such business.

They found competition from unexpected corners of industry. There was no real reason, as one historian explained it, "why the military should use them as prime contractors for its missiles rather than chemical companies like Thiokol or electronics companies like General Electric or Western Electric or automobile companies like Chrysler."[16]

Everybody seemed to want a piece of the space action. RCA was hard at work on a weather satellite; Ford's space division was on the verge of landing the contract to build a moon-impact capsule; AT&T's Bell Laboratories would soon begin work on the first communications satellite. Tiny Ball Brothers Research had its pointing controls and sun-seeking satellite.[17]

Still, the aviation giants emerged dominant. Martin landed the Navy's Viking and Vanguard rockets as well as the Air Force's Titan. Douglas got the Delta rocket. Lockheed won the Navy's Polaris submarine-based-missile contract and the Agena rocket booster serving as the backbone of the pioneering Discoverer/Corona spy satellites, also built by Lockheed.[18]

These were huge companies doing volume production using management systems shaped by strong military influence during the war years. At Martin in Denver, the environment was hierarchical, bureaucratic, and somewhat martial—quite the opposite of the university-reared Ball Brothers Research.

The difference was night and day, said Woolley. He was working on a testing program for Titan rockets in 1958 when a Martin manager suddenly banned coffee breaks. They were a waste of company time, said a man oblivious to the productivity gains caffeine brought to his little corner of the military-industrial complex. Drinking anything at one's workstation was forbidden.

Woolley declared such rules "silly" and drank Pepsi from a thermos bottle at his desk. His slight, erudite impression concealed an irreverence for rank, a product of his distaste for taking orders from "some idiot like my gunnery sergeant" in the U.S. Navy.[19]

Others in Martin's Titan operations acted in kind, leading to the bizarre scene, Woolley recalled, of grown men stealing sips of coffee from thermoses hidden inside their desks.

Woolley's manager let it slide. But one afternoon in 1957, his manager's boss was walking the aisles and, noting Woolley's thermos, stopped. "What's that?" the man asked.

"It's Pepsi Cola," Woolley answered.

"Get rid of it."

"That's what I'm doing," Woolley said, and took a drink.

It cost him his job.

Woolley went to work for Stanley Aviation at Denver's Stapleton Airport, which did big business in ejector seats for fighter planes.

Two years later, in the summer of 1959, David Stacey, Ball Brothers Research's technical director, called Woolley and asked him to work on their first satellite. Stacey had taken graduate classes with Woolley at the University of Colorado years earlier. Personal connections—in addition to engineering competence, often as demonstrated through candidates' working equations on chalkboards during interviews—were a big factor in Stacey's hiring decisions.

Woolley, 34 years old, told Stacey yes, he would be interested, but he was content at Stanley for the time being.

A week later, Woolley was fired. He had refused to dismiss a technician who had had the poor judgment to disagree with Robert Stanley himself. Stanley was an aviation giant. He had piloted America's first jet airplane, and then led the team that designed the first supersonic aircraft, the Bell X-1 that Chuck Yeager had flown to fame. But as Woolley put it, Stanley was "a brilliant, brilliant engineer and a nasty sonofabitch."

Woolley had closed on a house in Denver and had two small children. His wife Jean suggested he call Stacey back. Woolley resisted bargaining from a position of weakness.

"Well, maybe that'll tell you something about whether you really want to work for these people," Jean said.

Stacey offered a fair salary and later gave Woolley a bit of advice the younger man would heed for the rest of his career: "When you've got a problem to solve—an engineering problem—and you find a solution, if you can possibly arrange to do otherwise, don't just go with that first solution," Stacey told him. "Try to find a second, quite different way to solve it. You will find that combinations of the two will lead you to the best answer."

At Ball, Woolley went to work for Fred Dolder on the nascent sun-seeking satellite program. Dolder asked Woolley to build the satellite's gas-jet propulsion system. He would be a team of one.

Wooley's work in missile ballistics, rocket propulsion, and aircraft components seemed poor preparation for such a task. He told Dolder as much. Dolder replied that nobody at Ball knew anything about gas-jet systems, but Woolley seemed to have the most experience in learning how to do things he knew nothing about.

The mandate was vague. Bartoe had specified two types of gas "bottles" on the satellite. One, in the center of the spacecraft, was to supply nozzles adjusting the pitch of the spacecraft so that its instrument-pointing sail faced the sun within 3 degrees of perpendicular. The other bottles, smaller and on the flanks, were to feed jets keeping the spacecraft's gyroscopic wheel spinning at 30 revolutions per minute. That was it.

The Orbiting Solar Observatory, by then known as OSO (those on the project pronounced it "oh-so"), had beefed up to 450 pounds as the Delta rocket's lifting capacity increased. The spacecraft's rotating wheel was 44 inches across. The pointed sail had grown substantially, from Bartoe's proposal of two small rectangles of solar cells covering one square foot to a 3.3-square-foot half-moon with solar cells capable of generating 27 watts. The cells charged a battery delivering about 16 watts. Control systems and electric servomotors would get 7 watts; instruments, 9 watts.[20] You could run four OSOs on what it took to light a reading lamp.

The weight on the spacecraft's sail worried Dolder and Bartoe. The wheel needed to be either wider or heavier to maintain its dominance over the sail. With instruments filling five of the nine compartments and batteries and control electronics the other four, making the wheel heavier was impractical. The wheel couldn't grow wider because the spacecraft was already going to fill the payload shroud of the Delta rocket.

Bartoe suggested adding arms to OSO's wheel, and three sprouted from the spacecraft. Each arm would hold a volleyball-sized gas bottle. At launch,

the two-foot-long protrusions would hang over the third-stage rocket beneath it. Once released in orbit, the spinning craft's centrifugal force would lift and lock the arms parallel to the wheel body, giving OSO a diameter of nearly eight feet, a substantial span for something just three feet tall. The satellite suddenly gave the impression of some long-extinct crab.

By this point, the lone Control Cells building had been joined by another building called Tech 1. The new building had a tile floor, cinderblock walls, workshops, and optics and electronics laboratories stocked with oscilloscopes, photometers, and other specialist tools. Smoking was the rule, women were secretaries, and crew-cuts bristled. The nearest neighbor was the Humane Society, which the spacecraft builders dubbed the "Collie Cooker." Engineers create beautiful things in uninspiring places, and Ball Brothers Research was no exception, at least until one walked out the door and looked west to the mountains.

In his Tech 1 workspace, Woolley sketched out his plan. It involved a high-pressure tank, tubing, valves to cut pressure to manageable levels, and nozzles to shoot gas into space. An empty gas tank would end the mission. The propellant had to be treated like water on a desert crossing.

Woolley considered different gases. Bartoe had proposed helium. After hydrogen, helium had the most power per ounce among the gases (hydrogen's explosive tendencies worked against it). But Woolley dismissed helium as better in theory than in practice. The advantages of helium's high specific impulse—the energy provided per unit of propellant—would be more than offset by the larger, heavier bottle required to contain it. He settled on nitrogen. Factoring in the weight of the containers, nitrogen packed more than twice the punch of helium, he calculated.[21] It was also noncorrosive, nonpoisonous (cyanide gas would have had an even better power-to-weight ratio), and easy to get.

Woolley chose a storage pressure of 3,000 pounds per square inch. It was the same pressure contained by a scuba tank, so off-the-shelf parts were easy to come by. He chose the White Sewing Machine Company to make the bottles out of fiberglass epoxy. A White engineer later confided that they had strength-tested a prototype by dropping it into a window well from the fourth floor.

Woolley had the machine shop make him a few test nozzles. His colleagues on the electronics team needed to know the nozzles' thrust in order to design the circuitry connecting the spacecraft's photoelectric sensors to Woolley's jets.

To test the nozzles, Wooley invented a system involving a vacuum bell jar and an empty 500-gallon propane tank. He pumped out the propane tank, approximating a vacuum, and opened a valve between the tank and the bell jar. The propane tank sucked out the bell jar's air, flushing nitrogen from the nozzles and maintaining a vacuum for several seconds. He established the precise relationship between the behavior of the system's valves and the thrust emitted by a nozzle.

In the process, Woolley sought occasional guidance from Dolder.

"I would go to Fred and outline a problem to him, something that I was doing, on some subject that he knew nothing about and I knew a lot about, and he would sit down and he would say Dick, I wish you'd go back and think through this part of it again. And I'd say well, O.K., if you say so. I thought he was nuts, but since he was right so many other times, I'd go back and think through this and find that I'd made a blunder in there," Woolley recalled. "Apparently he spotted these things through something in my manner. Or I subconsciously knew there was something wrong there and he picked up on it. I don't know."

Help also came from other quarters. The one-time lifting and locking of OSO's protruding arms meant tubes channeling pressurized gas would have to bend. Woolley was considering swivel joints, which would be prone to leaking. Myron "Red" Poyer, a University of Colorado Rocket Project veteran who led the mechanical design work on OSO, went home one evening and screwed a pair of two-by-four boards to a door hinge, stapled copper refrigerator tubing along both arms and across the joint, and opened and closed his improvised clapper a hundred times. The Orbiting Solar Observatory would fly with refrigerator tubing.

PROBLEMS NEW TO ENGINEERING cropped up throughout the effort, and a team that grew to 40 people solved them using everything from rigorous mathematical analysis to plain common sense.

Related to Woolley's gas-jet work was the effort to understand what forces might poke at the orbiting spacecraft to change its orientation. Micrometeorites no heavier than flakes of cigarette ash could be ignored, the team decided. But they calculated out such effects as the difference in gravity's pull on the parts of the spacecraft farther and nearer the center of the Earth; collisions with the few air molecules still hanging on 300 miles into space; the magnetic pull of the planet on the spacecraft; and even the difference in pressure from the sun's rays on the sunward and leeward sides of the spacecraft. The combined influences of such forces would dictate how much gas Woolley's system would use and how long OSO could survive.[22]

The spacecraft's communications system proved vexing. Among the newcomers to Boulder was Reuben H. "Gabe" Gablehouse, descended from German beet farmers in Berthoud, 30 miles northeast of Boulder. Stacey had poached Gablehouse, a University of Colorado graduate who was well-versed in the emerging field of solid-state electronics, from Sandia Corp. in Albuquerque, New Mexico. Gablehouse eventually lured several colleagues to Boulder.

Prior to OSO, satellites either captured information on film or, more commonly, transmitted data back to Earth as instruments captured them. But that only worked when in range of a ground station, which constituted the minority of a given orbit. OSO was to record information from its instruments and its own status sensors for most of the orbit, then play it all back quickly while over a ground station.

The only way to store data was on magnetic tape. The team searched for someone willing to make miniaturized recorders capable of surviving spaceflight. Finally, Raymond Engineering Laboratory of Middletown, Connecticut, agreed to build two hermetically sealed, continuous-loop tape recorders. The 300-foot tape would record for 90 minutes and then play it all back during the five minutes the orbiter was within about 1,400 miles of a ground station. Flutter—the result of subtly changing tape speed across the read/write head—had to be countervailed both via special mechanisms in the recorders themselves and custom systems listening to the playback signal on the ground. The recorders used less than a watt of power and cost $33,000 each—more than 10 times the price of a new car. In the end, they were so difficult to make that Raymond Engineering would lose twice that amount on the deal.[23]

Transmitting the data also posed problems. The Ball team soon realized that beaming back information from a thousand miles away would take more power than off-the-shelf semiconductors could handle. Mercure called John Irwin, a solid-state physicist in the device department of Bell Laboratories, where transistors had been invented. They had been PhD students together at CU.

"I said hey, John, do you know anyplace where we could get germanium power transistors? Because we've got to figure out how we can get enough power to get this damn stuff down to the ground," Mercure recalled. "I didn't know what he was doing, but I knew he was in solid-state devices there. And he said, 'Yeah, we've got those. I'll just send you a handful.' He just gave them to us. Through the mail they came."

The mailman could offer no help with temperature control in orbit. Depending on whether the sun shined on a given part of the spacecraft or not, it might be bitterly cold or baking hot. There was no atmosphere to fan heat off sunny parts of the spacecraft, nor would OSO have the electrical or computing capacity to handle active heaters or air-conditioning units smart enough to turn themselves on and off at the right time. Their system would have to be passive, reflecting or absorbing just the right amount of sunlight, conducting just the right amount of heat.

Although everything on the spacecraft had its comfort zone, the nickel cadmium batteries were the most temperature-sensitive. They would work fine in a range that covers most earthly temperatures, 14 degrees to 95 degrees Fahrenheit. If the batteries got colder or hotter than that, they could fail or even explode.

Maintaining such a comfort zone for months in orbit had never been attempted. The team painted the inside of the rotating wheel black so heat could move evenly about the nine internal compartments and not pool in hot spots. Woolley, also responsible for the spacecraft's power system, encased the batteries in aluminum so they, too, could spread their heat throughout the wheel. But figuring out how much heat the sun, the sun-reflecting Earth, and the spacecraft's own innards would impart and how to balance temperatures was complicated.

Ball engineers knew their satellite would need some combination of sun-absorbing dark and sun-reflecting, highly polished surfaces. To understand how many square inches of either might be needed and where, they developed computer models of the spacecraft's shape and reflectivity to be run on a 950-pound, vacuum-tube-filled computer. [24] The models showed that the best answer was to develop special paints for the back of the satellite's sail and wheel rim, and to leave the top and bottom of the spacecraft's polished aluminum wheel alone.

The most vexing problem, though, was lubrication. Keeping OSO alive longer than three weeks became the job of Marion Fulk.

At 39, Fulk was an old man at Ball Brothers Research. Heavyset and with a receding hairline, he wore a jacket made of a Navajo rug. Fulk had grown up in Indiana and Illinois and began working in munitions as a civilian for the army, "mostly big bombs," during World War II.[25]

At the University of Chicago, he took classes in philosophy and studied such topics as how fine particles irritate lungs. He moved on to the University of Minnesota, where in the physics department he studied neutrons with scientists who had helped develop the atomic bomb. Fulk never earned a university degree. Robert Oppenheimer, who had led the Manhattan Project, once told him, "Don't worry. They'll think you have a doctorate anyway."[26]

Fulk left Minnesota for the Scripps Institute of Oceanography in La Jolla, California, developing equipment to detect tsunamis as they crossed the Pacific. "You could measure the depth of the ocean without getting your feet wet," he recalled. Fulk came to Boulder in the early 1950s to work in cryogenics on the same liquid-hydrogen-bomb project Mercure had worked on. Stacey hired him in 1959.

Fulk's resume spoke little of lubrication. He simply saw a problem that was "interesting and needed to be solved. And so we took out after it."

"We" was Fulk. He began by learning everything he could about friction. Fulk considered the phenomenon of cold welding in a vacuum as metallic confusion, something that happens when, "if you get two metals, when they touch, they don't know but that they belong together. They become coupled, and when they're pulled apart, they tear apart."

His nuclear-physics background was of no more help than his work in tsunamis. Nuclear forces might have been the most powerful in physics, but they tended to stay locked in the nucleus. Fulk needed to understand the behavior of electrons, which determine how materials combine and interact. He considered quantum electrodynamics, electric fields, induced fields, and other forces. Van der Walls, London, Casimir and Polder—all who had theorized about strange atomic attractions—became kindred spirits. He quietly became a top expert on the friction such forces create.

For Fulk, materials engineering began with theoretical mastery, the same progression Bartoe had followed with the science of gyroscopes and Woolley with propulsion gases. The same was true for those developing the satellite's antennas, its servomechanisms—even its paint. Nothing had been mandated by corporate policy. This was how simply the people Stacey hired happened to go about their problem solving, and it became one of the program's lasting legacies.

To speed the develoment of a lubricant that could survive for months in the high-vacuum, radiation-drenched space environment, Ball acquired vacuum chambers and hired colleagues for Fulk. Fulk's team ruled out Earth's slickest substances. Graphite, for example, depended on water molecules poached from the atmosphere, without which it crumbled. Other lubricants depended on contamination from the air.

What Fulk and his colleagues came up with remains a trade secret to this day, sold for use in military and commercial aircraft, medical and computer hardware, and, of course, spacecraft. It lubed the wheels of the Apollo moon rover and the innards of Apollo astronaut cameras. It was deemed too proprietary to patent, like the Coca-Cola formula.

"Vac Kote," as it was named, created a molecular bond with metals on moving parts and in the electric motors responsible for keeping the satellite spinning. It evaporated more than a million times slower than conventional lubricants, vital to both longevity and keeping clouds of grease from fogging the lenses of sensitive instruments. To the outside world, Ball Brothers Research described it as an "artificial two-dimensional atmosphere on moving surfaces," which reporters proved eager to repeat. With Bartoe's spacecraft design, it was one of the Orbiting Solar Observatory's great technical achievements.

At the same time, Ball Brothers Research was becoming a commercial success, ranking among NASA's top-25 contractors.[27] The bosses in Muncie had grown fond of their Boulder subsidiary, enough so that Ed Ball began speaking publicly about it. In late 1960, he explained to an audience of executives that "a scientist who can probe the secrets of the sun's burning energies may be a valued consultant on problems of heat transfer in a glass furnace."[28]

The Muncie parent's R&D chief Art Gaiser, initially dubious of the Boulder operation's reliance on government work, had come to believe that "scientific space projects will increase in number," underscoring the words, *"This is where our business will be derived."*[29]

Much of the work to conquer a collection of unknowns and combine the results into a functioning spacecraft could be described as applied science; indeed, OSO was advancing science—engineering science—before it ever left the ground. Such work was a means to an end, however, and the end was not coming nearly as quickly as the Ball team had promised. Fortunately, John Lindsay was on their side.

Chapter 7

Into Orbit

The Ball Brothers Research team had addressed stabilization and lubrication, the Orbiting Solar Observatory's most threatening technological demons. But lesser challenges proliferated, and it was clear that the idea of delivering three satellites in a year had been delusional.

The company made progress on two spacecraft. One was a fully developed prototype, which would bear the brunt of extreme testing. The second was the flight model, which, destined for space, would be treated with more respect. The third spacecraft, a second flight model, did not yet exist. Costs soared as months passed, with Ball Brothers Research's payroll expanding to 100 employees by early 1960.

The business of space is about balancing three factors: cost, schedule and risk. Risk mostly has to do with reliability. Spending more money and taking one's time designing, building and testing makes things more reliable; cutting costs and trimming schedules, the opposite. But after a certain point, spending more time and money mitigates less and less risk. Finding the right balance is the trick. In the dawn of the space age, cost and schedule were lesser concerns. "We've waited two thousand years to get this data," Harry Goett, the Goddard Space Flight Center's first director, argued to NASA headquarters. "We can wait another six months to get it right."[1]

OSO was costing NASA and American taxpayers far more than expected, but it could have been worse. Overtime was unpaid. Teams at Ball, Goddard and elsewhere labored through late nights and weekends. Fred Dolder worked seven days a week, coming home for a quick dinner with his wife and two young boys and going back to the office. Barbara Dolder asked about a family vacation. He told her he'd have to take a day off so they could talk about it.[2]

The Ball Brothers Research team was driven by something beyond money, and it wasn't necessarily patriotism either, despite the space race. "We weren't trying to beat the Russians or anything," Dolder recalled. "We were just trying to make something that worked."[3]

Just trying to make something that worked may have seemed a modest motivation for participants in such an audacious endeavor. But many of the Ball engineers hailed from rural Colorado and were the children of farmers, mechanics and the Great Depression. Tim Ostwald, among those who joined Ball to work on OSO, was one such man. He had grown up on a farm near Durango without electricity or a telephone. Among his boyhood vocations was ramming a horse-drawn sod plow through the clays of southwestern Colorado.

"We finally got indoor plumbing when I was in high school. I had barely been out of the county when I came to the University of Colorado," Ostwald said. "I made such a big step in my life that I think it was a bigger step than I could accommodate to become a space fan."[4]

Others saw something more transcendent in their quest. Merc Mercure, the young Ball Brothers Research president from the farming town of Loveland, had written in his PhD dissertation that space was "an area of scientific endeavor in which the smallest success brings with it a great personal sense of achievement and pleasure. There is a sense of adventure and no little thrill in being able to construct instruments to make measurements in a region where no man has ever ascended."[5] He later likened Ball's work to that of the great nineteenth-century English instrument makers, whose brass-sculpted masterpieces enabled the breakthroughs of scientists like Michael Faraday and James Clerk Maxwell.[6]

Things were changing fast in east Boulder. By late 1960, Ball Brothers Research had doubled again to 200 employees. In a report on how the company he co-founded was dealing with its explosive growth, David Stacey drew a graph showing the increase in employees over time. Ever the scientist, he used a logarithmic scale, "giving a straight line which can then be extrapolated." The straight line shot to the top of the paper.

Even in a company with half as many workers, Stacey said, "No longer can the management keep personal track of what's going on." Unwritten rules must give way to documented procedures, he continued, and better planning was critical with respect to buildings, people and equipment. "Almost never have we ordered machine tools before we needed them desperately," he admonished.

Stacey worried about the risk of relying on a single NASA contract for so much of the company's revenue. "To achieve permanence we must diversify. It is sure death to concentrate on one large program which can be canceled at any time."[7]

•

JOHN LINDSAY SHIELDED his Orbiting Solar Observatory from threats of cancellation. It was shaping up to be the most expensive solar physics experiment in history. But its costs paled in comparison to those of Goddard's Orbiting Geophysical Observatory and Orbiting Astronomical Observatory satellite programs. Even as OSO sucked down money, Lindsay's baby seemed a relative bargain.

It also helped that NASA was less bureaucratic than it would become.

"You didn't have to have a paper trail ten inches deep to do anything," said Nancy Grace Roman, the NASA headquarters astronomer who oversaw Lindsay's program. "It was relatively easy to get approvals because we recognized that we were doing something new and we had to learn how to do it. You could try things, and if they didn't work, you could try something else. I have in the past compared the beginning of the space program to a baby learning to walk. It falls down occasionally."[8]

The failures of the first six Project Ranger moon-probe missions in the early 1960s dampened NASA's and its political masters' appetites for risk. Failure after Ranger failure—each unique, each triggered by obscure oversights and mistakes—would focus the agency and U.S. Congress's attention on space program management practices, and the bureaucracy of full-on systems management became a necessary burden for all who followed.[9] The builders of John Lindsay's first satellite had the good fortune of avoiding such scrutiny. He made sure Ball had what it needed to build his sun-seeking satellite, and with minimal red tape or distraction.

"Lindsay would come out once a month. We didn't write progress reports, other than one-pagers. But for the real information that got transferred, Lindsay would come out here, and he would spend two, or three or four days at a time, and was intimately involved with what was going on," Dolder said. "We had an open relationship with John. We would say, 'Gee, we're in real trouble in this thing, or that thing.' And we never felt like we had to cover anything up, because his reaction was always one of trying to figure out how to help get the problem solved—not one of pointing a finger, or making us seem in a bad light."[10]

As costs mounted, Ball Brothers asked Lindsay for more money. Working with NASA's contract people, Lindsay sent more than Ball had asked for, which still turned out to be too little.

Upon one such request, Lindsay asked: "Don't you guys have some money reserved for the second flight model?"

He was talking about the third spacecraft specified in the contract. The first two, the actual Orbiting Solar Observatory and the prototype being abused during testing, were well along. Ball had yet to start working on the third, which was slated to join the first OSO in orbit. They had set money aside for it, though.

"Why don't you use that?" Lindsay suggested, thereby erasing an entire NASA satellite without so much as a memo.[11] He gave the project more time as well, pushing back the launch date a year from the original goal of late 1960.

Testing led to more testing. At previous companies, Tim Ostwald had followed formal procedures involving checklists agreed upon in advance by the customer, in this case NASA. He found no such system at Ball.

Testing was part of an engineer's, mechanic's or technician's job. They designed it. They built it. They would make sure it worked. Sure, there were test plans, but they were mostly in guys' heads. Dolder had few procedural safety nets or documents confirming the spacecraft had gone through trials. Lindsay didn't ask for proof. Project leaders trusted that Dick Woolley's gas jets, Marion Fulk's lubrication, and Pete Bartoe's dynamics would work.

"All the decisions were made between 5 and 7 o'clock in the evening. These meetings were in Dolder's office, with most of the people standing up. They'd decide what the next action would be if something was broken," Ostwald remembered.

Standard practice was to test the prototype at one-and-a-half times the violence expected during flight. Connections loosened, tubes broke, leads snapped. Solar cells on the spacecraft's sail peeled away during vibration testing and couldn't be salvaged, a $50,000 loss.[12]

The number of things to test was staggering. Even the spacecraft's magnetic tendencies had to be understood and, by placing the spacecraft structure in a large coil at different angles, neutralized. Otherwise the great magnet of Earth would pull the spacecraft's aim from the sun, wasting OSO's precious nitrogen fuel as it puffed to reorient. In simulations, the satellite used too much nitrogen to find the sun during the nine-second dawns and dusks of its orbit. Engineers added a set of "on-off" photoelectric sensors to keep the spacecraft from seeking its target until broad daylight. To test the spacecraft's communications, NASA had a U.S. Coast Guard helicopter dangle a model OSO a mile over the Fort Myers, Florida, tracking station. Engineers muffled the test spacecraft's radio to reflect the 300-to-1 difference with orbit altitude.

Problems proliferated. Tape recorders malfunctioned, aluminum castings came in flawed, scientific instruments broke. As Ostwald explained it, "Things just stack up one thing right after the other. You have a problem, so you work on that. Pretty far downstream you find out that something doesn't work as

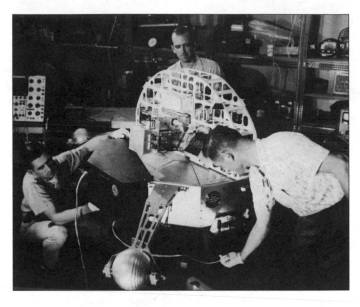

Stacey Johnson, left, Bob VonEschen and Jim Davis, OSO 1 launch team members, work on the OSO 1 spacecraft.

planned, so that pushes it out farther. Then the experimenters, a lot of times they were pushing the frontiers, doing things for the first time. And so the experiments weren't ready."

The University of Colorado experiment never would be ready. A machining problem grounded William Rense's ultraviolet spectrograph. Rense's technicians carefully aligned the instrument for testing, only to find the device out of focus days later. The metal holding the detector's optics had been improperly treated, shifting as it released internal stresses, much as a new house settles. Just months before the planned launch, Rense gave up.

Lindsay suddenly had a vacancy in the coveted real estate of OSO's sail. "They basically said, well, what can we put into this thing?" recalled Werner Neupert, a former Goddard scientist who helped Lindsay add a replacement instrument focusing on ultraviolet and X-rays.[13]

Even if Ball Brothers had known what they were getting themselves into, they probably would have bid OSO below its final $2.3 million price tag. That didn't include $2.5 million for the 13 instruments and another $2.5 million for the rocket.[14]

"Everybody bids optimistically on the schedule. Because that's part of the competition," Ostwald said. "They're optimists, thankfully, because that's the way you're most comfortable and that's the way you show your confidence and competence. People don't want to hear the truth. Especially managers and politicians do not want to hear the truth."

Launch was pushed back again, to early 1962, when the Orbiting Solar Observatory was shipped to Cape Canaveral in a giant can. A Ball Brothers Research contingent headed to the Florida coast, staying at the Surf Studio and Blue Horizon motels in Cocoa Beach and preparing for launch. Hierarchy evaporated entirely. Company president Mercure was handing wrenches to technicians. When they needed to change the balance weights on the sail's instruments, the Ball team looked for a machine shop in Cocoa Beach. They found one—a medieval-looking place with ancient machines powered by belts connected to a single motor—and convinced the owner to let a Ball machinist make new weights.

Days before the Thor Delta was to depart, Ball took apart the spacecraft to replace a flawed part in the communications system and plug a propulsion system leak. Neupert walked in and saw OSO parts scattered as if launch were a year away. The spacecraft was in the nose of the rocket when the team realized a connector still needed connecting. Woolley, among the slightest of the Ball group, was held horizontally aloft by colleagues, who guided his torso through a small door into the rocket shroud.

ON MARCH 7, 1962, at 11:06 a.m., a Thor Delta rocket carried NASA's Orbiting Solar Observatory to a 350-mile orbit. Ninety-six minutes later, the spacecraft radioed its first solar data. OSO was working.

Fifteen times a day, the satellite emerged from Earth's shadow. Its lubricated motors despun the sail, taking 45 seconds to lock onto the sun, with an accuracy of one minute of arc—thirty times better than a Rocket Project pointing control. OSO could aim through a basketball hoop from a mile away.[15]

Ed Ball, who had flown to Cape Canaveral for the launch, was so tickled he offered the Ball Brothers Research team members rides in his airplane. It was a triumph for the man whose gamble in Boulder had so vividly paid off. It was also a welcome distraction from home, where he was shutting down the antiquated Muncie glass plant his father and uncles had built 80 years before. It was now one of several Ball factories, and the least efficient.

Hugh Dryden, NASA's Deputy Administrator, declared the OSO spacecraft "[i]n many respects one of the most advanced satellites ever launched."

"With OSO," said Dryden, "we are beginning to probe deeply into the basic forces that determine the kind of planet we live on."[16]

Time called the spacecraft "a gadget-lover's dream"; *Popular Science* referred to it as "a dilly, a marvelously intricate piece of scientific machinery that is

The first Orbiting Solar Observatory lifts off on March 7, 1962.

performing like a Ferrari at Le Mans."[17] The *New York Times* ran a short story on its front page.

The local press took note of the home team's unlikely origins. "Many Coloradans were surprised to learn that the Orbiting Solar Observatory satellite launched from Cape Canaveral March 7 was built in Boulder by Ball Brothers Research Corp., prime contractor on the OSO project," the *Denver Post* reported. "But the name is familiar. Every housewife knows about Ball mason jars and lids, used all over America for home canning since before grandma was a bride."[18] Lindsay won the NASA Medal for Exceptional Scientific Achievement, the agency's top science award.[19]

Yet, as it had been with the Rocket Project's pioneering success, the press's attention span would be brief—a fact exacerbated by OSO's having launched two weeks after John Glenn became America's first orbiting human.

"That the American public now has acquired a taste for only the most spectacular of space feats is illustrated by the relatively little attention paid last week to the launching from Canaveral of one of the space age's most useful and sophisticated payloads: the orbiting solar observatory," *Science Magazine* wrote. "OSO, according to early reports, is performing satisfactorily, a matter-of-fact way of saying that it is an astonishingly successful piece of equipment. Some NASA engineers say the Mercury capsule is a Tinker Toy device compared with OSO, but, for understandable reasons, unmanned space feats now make poor box office."[20]

THE ORBITING SOLAR OBSERVATORY'S RIDE had its bumps. The most disturbing one cropped up as the accolades rolled in.

The observatory communicated with ground stations via a secret AM radio band. The radio system let the satellite's handlers command the spacecraft using seven tones capable of issuing 10 orders to the observatory. Switching the wheel experiments on and off consumed 20 percent of its vocabulary.

Soon after launch, Lindsay's team was alarmed to find the spacecraft, which had worked perfectly as it dipped beyond the eastern horizon, reappearing to the west in a disheveled state. Its experiments were off when they should have been on, and the tape recorder was playing back at high speed rather than recording new data.

"It turns out that almost any clever individual could probably trigger this satellite," Lindsay admitted. "It can be accidentally triggered by two radio stations that are approximately on the same frequency."[21]

Lindsay's team never determined what was meddling with their spacecraft. In April 1962, NASA dispatched 1,450 pounds of radio equipment and two Bendix Radio contract employees to Wheelus Air Force Base in Libya.[22] The setup shouted over the mysterious interference to correct the problem, but the damage to the spacecraft's tape recorder was done. All the extra stops and starts wore it out 25 days after launch. The backup recorder gave out in mid-May 1962.[23] From that point, OSO could send home only the real-time data it collected the few minutes each orbit during which it flew over tracking stations in Florida, Peru, Ecuador and Chile.

There were other problems. A week after the second tape recorder failed, a bad command from photoelectric eyes led Dick Woolley's gas-jet system to speed the wheel's spin rate to 50 revolutions per minute, 66 percent too fast. The electric servomotor counter-revolving the sail so it faced the sun was overmatched and shut down rather than draining the batteries.

The spacecraft's wheel had brakes, but they were built for one-time use— to slow the craft's rotation from the 120-rpm spin of launch to its 30-rpm operating pace. Once in orbit, friction and magnetic tugs from Earth would only slow down the wheel, engineers had believed. It was a month before the spacecraft's sail faced the sun again.[24]

Then came the beginning of the end. On July 9, 1962, the U.S. Air Force detonated Starfish Prime, a 1.4 megaton nuclear device, 248 miles above Johnston Island in the central Pacific. The test treated residents of Honolulu, Hawaii, 800 miles away, to a spectacular aurora. It was at least some compensation for the blast's electromagnetic pulse knocking out their streetlights and blowing their fuses. The Air Force declared the experiment a success. But the explosion's radiation-belt afterglow degraded the solar panels on OSO and other satellites—collateral damage in the colliding nuclear and space ages.

Thereafter, OSO was turned on at the beginning of a ground-station pass and turned off as it flew out of range to conserve electricity. That continued for two years, until May 1964 when the solar cells gave out.[25] The Orbiting Solar Observatory became man-made orbiting object 1962-ZETA 1. It tumbled for 17 years until plunging home as a ball of fire on October 8, 1981.

RATHER THAN BUILDING three satellites in one year, Ball Brothers Research had built one satellite in three years, for triple their original estimate.

But OSO delivered science beyond expectations, too. By the time its second tape recorder failed 77 days after launch, the satellite had observed the sun for 1,000 hours. It had watched 140 solar flares, mapped the sky in gamma radiation, and examined the low-altitude Van Allen Belt region. It found the sun's behavior as observed from space to be markedly different than what scientists had assumed from ground-based observations. It transmitted solar spectra that sent theorists back to their drawing boards to understand what sorts of atoms and molecules glow in what ways amid the sun's cauldron. The data it gathered in 11 weeks would have taken 12,000 sounding rocket shots to amass. The Orbiting Solar Observatory had brought home more information about the sun than had been collected in all preceding human history.[26]

Ball Brothers Research had grown to 365 employees. Many were hard at work on the follow-up OSO B spacecraft, a twin of the original, with technological improvements reflecting the three years that had passed since the first was designed. There was also a study contract for a much larger "Advanced OSO" satellite that Lindsay, his stature elevated by OSO's success, was promoting.

A *Business Week* reporter visiting the Boulder campus wrote that "Ball employees, most of them in their early 30s, putter about the premises in frayed tennis shoes and khakis."[27]

"There's a myth in this aerospace business that bigness is goodness," the 31-year-old Mercure told the magazine. "Hell, look at your big companies. They'll assign ten men [administering] to keep one man busy. We don't do it that way. We've got a very low percentage of paper-pushers; that's how we compete."

It was youthful hubris, and it met tragedy. The OSO B's travels ended in a 3,000-square-foot testing facility at the Kennedy Space Center. On April 14, 1964, Ball technician Lot Gabel stepped up to adjust a plastic bag shrouding the spacecraft and the third-stage rocket—a five-foot column of bottled fury—to which it was mounted. This innocuous act, like straightening a spouse's collar, triggered a spark of static electricity.

A design flaw caused the rocket engine to ignite. It smashed the spacecraft against the hangar's roof and then careened about the building, spraying blazing rocket fuel and finally ramming into a corner to burn itself out. Gabel, his Ball colleague Sid Dagle, and John Fassett of NASA died of severe burns. The 1967 Apollo 1 disaster, much more notorious, was no more lethal.

The diversification David Stacey had urged three years earlier had not happened. If the catastrophe were to have ended the OSO program, it could have ended Ball Brothers Research.

Compounding matters was a Government Accountability Office report taking exception to Lindsay's cowboy management style. The GAO castigated NASA for hand-picking Ball Brothers and said the program wasted $799,000 through extra effort caused by unrealistic delivery dates, "instances of poor coordination between technical and administrative personnel," and the costs of a "back-up spacecraft" (which, the GAO failed to note, had never been built).[28]

Lindsay's influence again proved decisive. Ball would have another shot at space, replacing destroyed instruments, repairing damaged hardware, and modifying the ill-fated spacecraft's prototype for flight. In Boulder, they worked through their mourning. The OSO 2 launched less than a year after the accident, on February 3, 1965.

It was John Lindsay's final orbiter. A malfunctioning rocket plunged the following OSO into the Atlantic in August 1965. A month later, on a Sunday afternoon in late September, the space scientist stepped outside his Springfield, Virginia, home to mow the lawn. After some minutes, the mower's back-and-forth hum fell into a steady drone. His wife looked out a window to see her husband face down in the grass, felled by a massive heart attack.[29] Lindsay was 48, with three young sons.

It's possible that his great accomplishments were behind him, and that his entrepreneurial style would have failed him in an increasingly bureaucratic

space agency. Regardless, John Lindsay's premature passing robbed the space pioneer of the historic stature he might otherwise have attained.

Lindsay had taken care of those who had taken him to orbit. Shortly before his death, he had secured Ball Brothers Research a contract for three more Orbiting Solar Observatories. The work served as the foundation from which the Boulder company, born in a University of Colorado basement, could develop into an enduring player in the space business.

Part II

Birth of a Mission

Chapter 8

From Sun to Stars and Back to Earth

Ball launched seven Orbiting Solar Observatories, the last of which flew in late 1971. The OSO 7's wheel was heavy enough to keep the gyroscope a gyroscope without the protruding arms, and a rectangular solar array tripled the first OSO's power. The satellite weighed three times as much as the first OSO, carrying more than the original's weight in instruments alone. The final OSO could point seven times more accurately, aim at specific spots on the sun, and even scan the entire disc like eyes reading from a page. Ball's last OSO carried a sun-occulting coronagraph—the instrument the CU Rocket Project proposed in 1948 but could not build. Ball had been flying such instruments since the mid-1960s. The impossible had become routine.

The OSO program ended bitterly. Ball had bid on a major contract for three follow-on spacecraft, known as OSO I, J and K, only to lose to Hughes Aircraft in December 1970. Hughes had promised more spacecraft for less money. Ball Brothers Research, and in particular its new president Pete Bartoe, said it couldn't be done for such a price.

For the first time, the Boulder company had to lay people off.[1] The loss was such a shock that U.S. Senator Gordon Allott of Colorado demanded an explanation. The NASA chief responded with a letter highlighting "significant weaknesses" in the Ball bid, saying, in effect, the progenitor of several Orbiting Solar Observatories was ill-equipped to produce the new OSOs, much heavier spacecraft with better pointing accuracy, more power and greater data capacity.[2] Gone was John Lindsay's great gift to Boulder, the cash cow driving the company's growth since 1959.

Hughes went way over budget, spending the entire contract on a single spacecraft, OSO 8, which launched in 1975. An annoyed NASA then killed the Orbiting Solar Observatory program. As a sort of OSO swan song, Ball built a similar satellite for the U.S. Air Force, the P78-1, which launched in 1979. It gathered solar data until September 1985, when a fighter-launched missile blew it from its 370-mile orbit, thereby giving the orbiter the distinction of being the first spacecraft ever taken out with an anti-satellite weapon.

Ball engineers in the satellite's control room watched their screens go dark with the hit. That Ball's "black" (classified) business had helped the Air Force develop the missile system was some consolation.

With the OSO program gone, Ball was fortunate to have developed a healthy, if less predictable, business in space instruments, for which the jar company's offspring would become best known.

Like most everything else at Ball Brothers Research, the instrument work had started with pointing controls. The company focused the bulk of its energies launching one OSO after the next. But a smaller team built 45 pointing controls throughout the late 1950s and 1960s. Each took months to construct and test. Part of the effort was to make sure the instruments supplied by outside scientists worked with the pointing control hardware aiming them at the sun. University professors and military scientists had a habit of arriving in Boulder with instruments that stood little chance of surviving launch, or with optics too dirty to see the sun in the ultraviolet, or with poorly conceived data transfer or radio schemes. Firms specializing in space instruments didn't exist.

"We just started helping out. First a few little electronics boxes here, then a connector, then an assembly there," recalled Bill Frank, a Ball engineer who had worked on pointing controls from the company's earliest days.

Soon they were building optical systems, then complete instruments. Ball's pointing-control engineers recognized that scientists often wanted things they could not articulate.. The team gained a reputation for accommodating them.

"If the scientists wanted the damn thing painted pink with polka dots, we'd paint it pink with polka dots. But we'd use good paint," Frank said.

They eased into satellite instruments. The first, a cosmic-ray experiment for the University of California, launched on the Orbiting Geophysical Observatory in 1964. The next flew on their own OSO 2. In the mid-1960s, Ball proposed several larger and more sophisticated solar instruments for John Lindsay's Advanced Solar Observatory. When that satellite disappeared in a 1966 NASA budget cut, Bartoe came up with a novel place to put the instruments.[3] He called it the Apollo Telescope Mount.

The Apollo astronauts' trip to the moon would take three days. Bartoe's idea was to put their commutes to good use studying the sun. A latticework arm with several instruments would rise out of the service module—a pointing control for Apollo as opposed to an Aerobee. Astronauts would collect and reload film. It was to happen by 1968, a peak in the eleven-year solar cycle.

The Marshall Space Flight Center got wind of the idea and morphed Bartoe's Apollo Telescope Mount into Skylab, a program of breathtaking ambition. It would launch in 1973, carrying four Ball solar-science instruments worth tens of millions of dollars—two for Naval Research Laboratory scientists, one for the High Altitude Observatory, and one for Harvard. The work helped Ball Brothers Research overcome the OSO loss.

By the late 1970s, Ball had harvested expertise in cryogenic fluids tracing back to the liquid hydrogen bomb to build the world's first cryogenic space telescope, the Infrared Astronomical Satellite. IRAS sensed infrared wavelengths coming from the far corners of the cosmos.

The expanding universe stretches light, shifting it toward the color red. The older the star or galaxy, the redder it looks. Infrared light—reds too stretched out for the human eye to see—is heat. The motivation behind IRAS was similar to John Lindsay's for an orbiting sun-seeker. Lindsay had to send his instruments above the ozone layer to see the ultraviolet; IRAS had to escape Earth's heat to see the infrared. But IRAS also had to keep itself extremely cold or be blinded by its own light. Ball developed a liquid-helium-cooled telescope to keep its optics at about 4 degrees (2 C) above absolute zero.

The telescope, which could sense the tiniest flicker from far-off stars and galaxies, opened up a whole new universe. Scanning 96 percent of the sky in the 10 months until its coolant bled away, IRAS spotted 350,000 infrared-emitting sources, doubling the number of cataloged astronomical objects. It discovered six comets for good measure, and gathered new information on a comet called Tempel 1, which would become the object of a bold Ball mission many years later.[4]

Ball's cryogenic skills later played into Goddard's Cosmic Background Explorer, or COBE, satellite. Launched in 1989, COBE spotted cosmic microwave background patterns to match those expected by the Big Bang theory of the universe's creation. The physicist Stephen Hawking called it "the discovery of the century, if not of all time."[5] COBE's lead scientists won the Nobel Prize in physics in 2006.

Starting in the late 1970s, the Hubble Space Telescope became a Ball focus. The company's 20-year relationship with NASA Goddard, which managed the Hubble program, paved the way. Ball engineers came to understand the space telescope as few did, and their Goddard High Resolution Spectrograph launched with Hubble in 1990. The company added instruments on subsequent shuttle-servicing missions, the most famous being COSTAR, the "eyeglasses" that helped fix Hubble's hazy vision.

MURK BOTTEMA, Ball's top optical engineer, designed COSTAR. Bottema was born in Velsen, the Netherlands, in 1923. As a young man during the waning days of World War II, he stayed off the Nazi-occupied streets during daylight hours for fear of being picked up and conscripted. He spent his time studying mathematics and physics and tinkering, at one point building a functioning clock with his Erector Set.

Bottema was an Old World gentleman before, technically, he was a man. When his family moved to the Dutch town of Assen, a girl of 15 looked out her window to see a boy step out of the back seat of the family car and open the door for his mother. "The way he looked, so intelligent and so polite, I said 'This is the man I'm going to marry,'" Willie Bottema recalled. And she did.

Her young husband went on to study physics under Nobel Prize winner Fritz Zernicke at the University of Groningen, where Bottema earned his PhD. During a year at Johns Hopkins University in 1958, he so impressed his hosts they asked him to return permanently. He emigrated with his wife and three young boys to work designing rocket-borne telescopes and spectrometers for the study of planetary atmospheres.

In 1968, Bottema joined Ball in Boulder. He was a quiet, serious man, with great confidence. As his wife put it, "He trusted his own brain."

When the plan for the Hubble Space Telescope solidified in the 1970s, Bottema worked with NASA scientists on the design and came to know Hubble intimately. He proposed several Ball instruments for the telescope, three of which Ball would build. He retired in 1990, thinking his contributions to the space telescope were behind him.

Hubble was a silver-sheathed cannon four stories tall and weighing more than 12 tons. Technical problems pushed back a planned 1983 launch; the space shuttle Challenger disaster in January 1986 delayed a liftoff slated for August of that year. When Hubble finally launched in April 1990, it was to be capable of discerning the period at the end of this sentence from a mile away. It would, NASA promised, bring home images with ten times the clarity of the best earthly telescopes.[6]

"In a thunderous overture to a promised new era in astronomy, the space shuttle Discovery rocketed into orbit today with the $1.5 billion Hubble Space Telescope, which scientists believe will give them a commanding view of the universe as it was, is and will be," a front-page *New York Times* story gushed.[7]

The tenor would change. Two months later, NASA announced that Perkin-Elmer Corporation, the mirror contractor, had ground the giant telescope's eight-foot primary mirror with slightly less curvature around its fringe than they should have.[8]

Hubble's flaw, caused by what's known as a spherical aberration, stretched the tip of the telescope's point of focus so much that there was no real focal point. The mirror's error was fantastically minor—roughly 1/25 of a hairs-breadth. Scaled up, the flaw would represent a difference less than the tip of a ballpoint pen across a football stadium. But with optics as precise as Hubble's, the imperfect mirror affected not only image sharpness, but also the telescope's ability to detect faint objects, which was to be one of Hubble's greatest strengths.

Ed Weiler had spent a decade of his life on the most expensive science experiment in history, and NASA's chief scientist for the space telescope understood it to be fundamentally flawed. He faced the press on June 27, 1990. "It would be dishonest to say the mood of the scientists is happy," he said. "We're all frustrated, obviously. But we should be able to fix it in the long run. Nobody's walking away."[9]

The public reaction was swift and brutal. "Pix Nixed as Hubble Sees Double," "Space Telescope Can't See Straight," "Hubble, Double, Toil and Trouble," shouted the headlines.[10] *Newsweek* put Hubble on its cover with the words: "Star Crossed: NASA's $1.5 billion blunder." U.S. Senator Barbara Mikulski of Maryland, a reliable space booster whose state hosted much of the NASA's Hubble effort, called the telescope a "techno-turkey." David Letterman piled on, the late-night comedian's top-ten list of NASA excuses for Hubble including "The guy at Sears promised it would work fine," and "Bum with squeegee smeared lens at red light."[11]

Yet, scientists and engineers realized, Hubble had the advantage of having been ground to perfect imperfection. Even before Weiler spoke publicly of Hubble's myopia, scientists suggested that future Hubble instruments be equipped with small mirrors intentionally shaped to cancel out the problem. As one Hubble scientist put it, "as applied to light waves anyway, two wrongs can make a right."[12] Jet Propulsion Laboratory engineers were already working on a second-generation replacement for Hubble's main camera called the Wide field and Planetary Camera 2. They figured they could add a nickel-sized correcting mirror.

That was fine for a future instrument, but Hubble was flying now, with hundreds of millions of dollars in science projects drinking in bad light. NASA asked some of the nation's top optical minds for ideas to salvage what was already in orbit. The Space Telescope Science Institute leading the Hubble astronomical effort formed the HST Strategy Panel, including famed astronomer Lyman Spitzer, astronaut Bruce McCandless and more than a dozen others. Bottema was among them.

The group vetted dozens of ideas. Scientists and engineers proposed wrenching the faulty mirror into the correct shape; sending an astronaut or, more likely, a robot spelunking into the school-bus-sized barrel of Hubble to replace the secondary mirror; recoating the edges of the flawed mirror in space; and bolting a massive optical corrector lens into the inlet of the telescope to reshape light before it entered.[13]

Bottema, coming out of retirement to tackle the Hubble problem, had another idea. He suggested using relay mirrors similar to those planned for JPL's next-generation main camera. The mirrors would pick off incoming light, remold it, and then bounce it to Hubble's scientific instruments. Just how one might arrange a bunch of coin-size mirrors inside the orbiting telescope came from an inspiration in a German hotel shower. While at a two-day Hubble Strategy Panel meeting in Garching, Germany in September 1990, James Crocker, a Space Telescope Science Institute engineer on the panel, was soaping up when he noted the adjustable, articulating arms holding the shower head in place. Inspiration struck. "I could see Murk Bottema's mirrors on the shower head," he later explained.[14]

Back home, Crocker built a mockup using foam board, his son's toy Ramagon construction set, and a few metal discs from the local hardware store. He presented his creation to colleagues at the strategy panel's final meeting in Baltimore. The idea was to sacrifice one of Hubble's four scientific instruments and replace it with a new device whose sole job was to fix the light entering the other three. Crocker was soon leading the team charged with creating such a thing.[15] Ball Aerospace, as Ball Brothers Research was now known, started the work of designing and building it in January 1991.

Bottema went to work. The Dutchman's depth in optics was akin to Pete Bartoe's skill in mechanical design, involving a comprehension that plumbed the depths of the underlying physics. They shared an ability to visualize in three dimensions and then convert their mental imagery into mathematics.

Bottema scratched out pages of integrals and partial derivatives and Bessel functions, sketching light paths darting from mirror to mirror. He used computers, but mainly to confirm what his mind and hand had already established. Bottema was among the last of a breed. Modern optical engineers do some algebra to set boundaries, then pour their rough outlines into optical-design software with names like Code V, Zemax and Oslo, which churn out a range of shapes and surfaces and positions and suggest optimal configurations. With computers, there is little need to understand optics the way Bottema did. He intuitively knew why a lens or mirror behaved in a certain way, how it related with all the other optical surfaces, and how it shaped the final image. A computer can arrange and optimize, but Bottema could observe a blurred image and calculate *why* it was blurred. Then he could work the problem forward again to figure out what sorts of optics would correct it.

Bottema spent his days on the Hubble problem, but his best hours were at night. After tea, reading the newspaper and a late dinner, he sat at the dining-room table with pencil and paper, writing equations. When things were going well, he whistled. When they went less well, he paced back and forth through the living and dining rooms. He talked through nettlesome technical details with his wife, who sewed on the couch under a window that looked out on the southerly portion of the Flatirons formation. She didn't understand him, knew he was answering his own questions, and stitched away.

He designed a pair of mirrors on a telescoping arm. The first would reach out and snatch Hubble's light before it could feed an instrument, redirecting it to a second mirror. The second mirror would be deformed in a way precisely opposite to the flaw of Hubble's primary mirror but 200 times smaller. He solved the equations and showed that a few strategically placed mirrors could deliver corrected light to three Hubble instruments.

This work provided the basis for the Corrective Optics Space Telescope Axial Replacement, or COSTAR, which a team of about 400 workers built at Ball Aerospace, in a crash 28-month program. From Crocker's vision in a German shower and Bottema's optical mastery evolved a $50 million master-piece of space hardware. Built into a spare Hubble instrument box the size of a voting booth, the business end of COSTAR would reach out into the body of the telescope itself. The optics package squeezed ten mirrors, four telescoping arms, a dozen electric motors and various heaters, wiring and sensors into a four-foot-long triangular prism with a cross section the size of a slice of apple pie.[16]

In December 1993, the space shuttle Endeavor made the first house call to Hubble. Astronauts replaced the main camera with JPL's new one fitted with internal "eyeglasses" and sacrificed a high-speed photometer to make room for COSTAR. When its mirrors extended, it was as if COSTAR's three neighboring instruments had undergone laser eye surgery. Aside from light lost along the way—an inevitable by-product of relay mirrors—the new JPL camera and COSTAR restored Hubble's vision remarkably. Less than six months later, scientists announced a COSTAR-corrected Hubble instrument had brought home hard spectral evidence of a supermassive black hole at the heart of a galaxy 50 million light years away, confirming the existence of the long-theorized cosmic vacuum cleaners once and for all.[17] Weiler considered it "the single biggest observation that COSTAR enabled. And it's slightly important, because it changed something from Star Trek fantasy into scientific reality."

Bottema had not lived to see the day. In July 1992, as his creation took shape, he succumbed to cancer. His Ball colleagues mounted a plaque in his

Shuttle astronauts prepare to install COSTAR during the first Hubble Space Telescope servicing mission in December 1993.

memory on COSTAR. Bottema had had no doubt the instrument carrying it would work.

"You don't have to worry," he had assured his wife.

BALL CONTINUED TO BUILD satellites, though that business kept a lower profile. One notable effort was the Solar Mesosphere Explorer (SME), a small spacecraft for understanding ozone creation and destruction in the upper reaches of the atmosphere. The satellite, launched in 1981, emerged through the reuniting of the old University of Colorado Rocket Project's parted halves. Ball built the spacecraft; CU's Laboratory for Atmospheric and Space Physics, as the Upper Air Laboratory was renamed in 1965, built the instruments and handled mission operations. Although SME blazed a trail in modern low-cost spaceflight, it failed to spark interest at NASA for similar missions, the trend at the time being toward massive spacecraft carrying an arsenal of instruments. SME would prove to be a decade ahead of its time.

The Earth Radiation Budget Satellite (ERBS), a larger satellite built to help scientists understand how the Earth absorbs and reflects the sun's energy, was the first designed for a space shuttle launch. ERBS was a bulky, odd-shaped creature designed to minimize the space it occupied in the shuttle bay. The satellite gamed the system: NASA charged by volume rather than by weight, a policy it quickly reconsidered.

As with nearly all satellites, ERBS had to be spin-balanced like the wheels of a car. Yet the satellite was too big to spin and so oddly shaped that air resistance would skew the results. They called in Dick Woolley. Woolley, who had designed the Orbiting Solar Observatory's propulsion and electrical systems, had spin-balanced Ball's first satellite and every one since. He had developed a reputation for coming up with simple solutions to vexing problems. "You'd think, 'This is never going to work,'" as a former colleague put it. "But it always did."[18] Woolley attributed his seemingly mystical abilities to a mastery of the basics. "I made my whole career out of knowing sophomore physics," he said. "I know it thoroughly, so I can solve problems many PhDs can't solve."[19]

For ERBS, Woolley created the Ball Dynamic Balancing Machine, Mark III. It became better known as the "Woolley Wobbler." It deduced a spacecraft's balance points by rocking it rather than spinning it. The Wobbler balanced ERBS and then waited in a warehouse until another oddly shaped spacecraft called on its services two decades later.

When Challenger carried ERBS to orbit in 1984, Sally Ride, America's first female astronaut, shook the outstretched shuttle arm to shoo Ball's creation away like an errant insect. It survived more than 20 years in orbit.

ERBS and SME differed from previous Ball creations in that they looked at Earth rather than the sun or the stars. The expertise Ball honed in designing and building these satellites brought follow-on Earth-observing work from NASA and foreign space agencies as well as from the military, for which Ball was building secret spacecraft and instruments.

Ball gained a reputation for achieving the impossible for exorbitant amounts of money. The work environment retained a campus feel and lacked the rigid hierarchy and politicized atmosphere of larger competitors. Perhaps Ball had its university heritage to thank; maybe the culture had its roots in Muncie, where Ed Ball's father, a man of wealth and prominence, had spent many days among hourly workers on the glass-factory floor. Or it could have been the percolation of the first OSO team throughout the ranks of the company, where their mutual respect trumped titles. Gablehouse, the company's president, wrote, "Having fun is mandatory" on blackboards before meetings. Whatever the reasons, Ball retained an informality rarely found in the regimented aerospace business.

The company diversified. By the mid-1980s, Ball was doing about 40 percent of its business for the military. A satellite system developed for the Reagan administration's Strategic Defense Initiative fired a laser beam onto a mirror on the bottom of a satellite moving five miles per second, 217 miles over Hawaii. The mirror reflected the laser light onto the bull's-eye of a target 12 miles away—for as long as 80 seconds straight. The Relay Mirror Experiment, as it was called, tested technologies that might help knock down an intercontinental ballistic missile, and it advanced spacecraft pointing accuracy by a factor of 100 or more. In Boulder, peace activists staged a two-hour protest; Ball learned to keep its defense work quiet.

Boulder had more than quadrupled in population, to 100,000, since the 1950s. The city had gone from conservative hamlet to the liberal bastion of the mountain west—the "People's Republic of Boulder," as detractors from beyond the "Boulder Bubble" labeled the place. Ball spawned technology startups and contributed to a tech-worker base IBM and others came to exploit and nurture. The town in which the arrival of a sketchy weighing-device company was once big news had evolved into a regional technology hub.

Defense work led to the QuickBird series of Earth-observing satellites. Using them, DigitalGlobe, a Ball offshoot, sold imagery capable of distinguishing a car from an SUV from 300 miles up. National-security types were big customers, and the satellites' imagery became a mainstay of popular mapping Web sites.

As the glass company's Boulder operation built dozens of other spacecraft and instruments, it took many side trips. At one time or another, Ball made mainframe computer memory cores, large-scale irrigation equipment, low-light cameras for spotting enemy ships at night, modular homes, a system Bartoe helped develop for sending deepwater TV images wirelessly to the surface, and a protective coating for vinyl LP records. Some sold, some didn't.

Experimentation with antennas lying flush against Aerobee rockets led to a major side business in microstrip antennas, which found their way onto nuclear missiles, fighter planes, commercial airliners, and Air Force One. The same technology went into billboard-sized antennas for microwave imaging of Earth and oceans, the technique's chief benefit being an ability to see through clouds.

PETE BARTOE'S SWAN SONG was one of Ball's most interesting forays. By 1973, it was clear to Bartoe as well as Ball that his skills as an executive did not—could not, really—equal his gifts as an engineer. Bartoe had never been a polished political creature. A newspaper account described him as "a hard-working, tough minded leader who is undoubtedly 'boss'" around Ball, a man who "speaks and acts with refreshing directness."[20]

"It seems to me he did a quite satisfactory job as the boss man in Boulder," Woolley said. "But he is a great engineer. And it seems a shame to waste a great engineer as a manager. It seems to me that managers are a dime a dozen in comparison. Now no manager would agree with that, I suppose."

The 45-year-old Bartoe convinced John Fisher, who had succeeded the retiring Ed Ball as the Muncie parent's president in 1970, to let him build a jet airplane. The idea was to blast jet wash over the aircraft's wings to harness something called the coanda effect, the same phenomenon that makes water cling to a glass held sideways under a spigot. He would produce a prototype short-takeoff-and-landing jet aircraft—the Jetwing, Bartoe called it—such as the world had never seen.

"Every big aircraft company has spent a few million bucks to try to find a way to increase the circulation around the wing to improve lift," said Woolley. "And here this guy comes along and says, 'I know how to do that.' But those of us who knew Pete were of the opinion that of course it would work. If Pete says he can do it, why, he can do it."

Ball-Bartoe Aircraft Corporation was born. Bartoe moved into a small office at the same Boulder airport where he had lived and worked as a college student and, later, flown his biplanes. He churned out drawing after drawing at a drafting board. Marv Williams, who had led the machine shops at the University of Colorado Rocket Project and later at Ball Brothers Research, built the jet with Brad Davenport, a fellow machinist. When Williams fell ill, his son Sig took his place. There was a procurement man named Ernie Malekowski. It was the only modern jet-aircraft company whose entire staff could gather in a restaurant booth. On the side, Bartoe designed, and his colleagues built, a propeller-driven aerobatic biplane called the Skyote. Small enough to fit into a two-car garage, it combined traits of the Bucker Jungmeister and the Rose Parrakeet planes Bartoe liked best. The idea was that enthusiasts would buy Skyote kits and build their own airplanes.

In 1977, Ball-Bartoe Aircraft's titanium-sheathed Jetwing roared into the sky. It could fly at speeds ranging from a jet-like 345 mph all the way down to 45 mph, which Model Ts were known to hit. Bartoe fancied the idea of landing backwards in a stiff headwind.

The Jetwing was as abject a commercial failure as it was a technological masterstroke. The Navy was more interested in vertical takeoff; the Air Force had plenty of runway room; passenger-aircraft builders liked the idea of short

Pete Bartoe and his
Jetwing

takeoff and landing, but were less enthusiastic about the lower cruising speeds
and higher fuel costs. Ball-Bartoe Aircraft quietly closed its doors.

Fisher donated its sole product to the University of Tennessee, his alma
mater, where it flew for a few years and then landed in a warehouse.[21] Its cre-
ator took to the seas.

Bartoe took a leave of absence and captained his 32-foot ketch from California through the South Seas to New Zealand. His wife and daughters joined
him for some of the journey. But much of the time he sailed alone, navigating
beneath the sun and the stars and the spacecraft of his design. In New Zealand,
an *Auckland Star* columnist profiled Bartoe under the title, "He's sailing away
from a jet wing."[22]

He was sailing away from Ball, too. "If you bob around on a small boat for
a couple of years, your perspective of what's important changes," Bartoe said.

There was nowhere for him to dock in the company he helped create anyway. Bartoe retired to a place he named Skyote Ranch in the woods of Clark,
Colorado, not far from Steamboat Springs. He built a log home, cleared an
airstrip, and never designed another flying machine.

ED BALL'S INVESTMENT in Boulder transformed the jar company his father
co-founded. As he had envisioned, Ball's aerospace business became a font of
talent for the parent company, providing engineers to solve production problems and managers to serve as top executives in Muncie. It lent marketing buzz
other container companies couldn't touch. In the early 1970s, Ball ran ads in
trade magazines with images of moon rovers and extraterrestrial scenes. "From
the Sea of Tranquility to your packaging department," began one. "The same

team at Ball who help NASA put men on the moon make better glass for you. Ball space program breakthroughs make Ball glass better."[23]

In 1969, Merc Mercure's Boulder operation bought Jeffco Manufacturing Co. in Golden, Colorado, about 20 miles south of Boulder. Jeffco Manufacturing made cans for the Coors brewery. It was the first of several aluminum-can plants Mercure would acquire for Ball.

Beer and soda cans became Ball's biggest business, with factories around the world producing billions of aluminum containers a year, later augmented by plastic bottles. The glass business faded until the Muncie company sold off the operation in 1993. One can still buy new Ball jars, but Jarden Home Brands makes them.

Ball's Colorado presence grew. In 1987, The company had bought a 216,000-square-foot building in Broomfield, Colorado, halfway between Denver and Boulder. Although the city welcomed Ball, Broomfield had really been pulling for Yugo, the Yugoslavian car maker, which had also shown interest in the property. "That would have been a bigger boost to the local economy," the head of the city's economic development office lamented.[24]

In 1998, with its Muncie manufacturing base gone, Ball Corp. pulled up stakes from its home of more than a century, moving its headquarters to the building in Broomfield. Ed Ball's ill-advised investment in a tiny Colorado weighing-device company had, 42 years later, uprooted the parent itself from "Middletown" in favor of something closer to the country's geographic mid-point. It was a dark day for Muncie, which had never recovered from Ed Ball's shuttering the local glass plant in 1962.

The aerospace business saw its own share of upheaval. Over the years, Ball Aerospace's competitors merged and consolidated into giants. Boeing swallowed up North American Aviation, McDonnell, Douglas and Rockwell. The Glenn L. Martin Company, where Woolley and other Ball Brothers Research employees had worked, joined with Lockheed. Northrop merged with Grumman. General Motors bought Hughes Aircraft. Only a few—Aerojet, TRW, upstart Orbital Sciences, Ball—retained their original character. Four decades after its founding, Ball Aerospace remained something of an aerospace-industry boutique competing against companies 40 times larger. Despite ongoing efforts to land another OSO-style gravy train, Ball continued to craft instruments and spacecraft one at a time—Fabergé eggs, where others sold them by the dozen.

The aerospace business had always been feast or famine, and the combination of economic recession and the "peace dividend" following the fall of the Soviet Union made for hungry times. Ball Aerospace laid off about

Difficult times in the early 1990s did not prevent Ball from thinking big. This
artist's rendering was part of a two-page advertisement with the tagline, "The next
time you need an innovative approach in solving your problems, think Ball."

1,000 staff between 1990 and 1994. By then, the company was down to 1,250
employees—roughly the size it had been 25 years earlier. Eleven of 19 Ball
Aerospace campus buildings in east Boulder were adorned with "for lease"
signs. Ball Corp. wavered, hiring an investment bank to investigate "strategic
alliances, joint venturing…selling all or parts and spinning off all or parts" of
its aerospace division.[25]

But the economy improved, and the work rolled back in. In part, the
turnaround hinged on a new NASA focus on "faster, better, cheaper" missions,
which favored those able to develop small spacecraft as well as the instruments
riding them. Ball was such a company. Among the most interesting of its prospects was a far-fetched comet mission.

Chapter 9

Discovery and Deep Impact

Throughout the 1980s, billion-dollar "gigabuck" missions dominated the American space program, with mixed results. On the manned side, there was the space shuttle, a tractor trailer to low Earth orbit estimated to cost $1 billion per delivery. In astronomy, taxpayers spent some $1.5 billion on the Hubble Space Telescope only to find its vision impaired. To study the planets, massive spacecraft such as the Galileo Jupiter orbiter and the Mars Observer sucked up more than $2 billion combined. In August 1993, Mars Observer went silent three days before reaching the Red Planet, the victim, as close as engineers could tell, of an exploding propulsion system. Galileo arrived at Jupiter in 1995 unable to unfurl its main antenna, which was to spread like a Japanese fan. The backup antenna, capable of transmitting just one image per month, was all that was left. Engineers managed to boost the backup's speed by a factor of 200, but it still left scientists with one-tenth of their expected harvest from the giant planet and its many moons.[1]

With the space shuttle ferrying nearly all U.S. science satellites into orbit, scientists and satellite builders had little incentive to cut size and weight. Then the 1986 Challenger disaster grounded the shuttle fleet, putting conventional rockets back in the space science business. Rockets came in many sizes. The smaller the satellite, the smaller the rocket to launch it, and the lower the cost.

These and other factors triggered a revival for smaller satellites. They had less capability, true. But they would cost less, and you could launch many more of them, with the fringe benefit of giving graduate students and young scientists something to work on between once-in-a-decade, swing-for-the-fences flagship missions.

NASA leadership began its push for smaller spacecraft in 1989, a strategic shift culminating in a 1991 conference in Woods Hole, Massachusetts. About sixty space scientists, "in a tense atmosphere that crackled with disagreement and open conflict," agreed to focus on more, smaller, cheaper, and simpler spacecraft.[2]

Congress followed suit. In 1992, legislators asked NASA for "a plan to stimulate and develop small planetary or other space science projects, emphasizing those which could be accomplished by the academic or research communities."[3]

NASA responded with the Discovery Program. Rather than a dozen instruments hanging like tree ornaments from a spacecraft, there would be perhaps two or three. Programs would go from concept to reality in three years. A space scientist with the title Principal Investigator would lead each mission, with a NASA center such as the Jet Propulsion Laboratory or the Goddard Space Flight Center acting as NASA's eyes on the ground. Small teams would build and test spacecraft without the raft of paperwork weighing down big missions, and NASA would give project leaders the freedom to get the job done as they saw fit. Missions would opt for off-the-shelf parts wherever possible. The fruits of high technology, in particular miniaturization, would also work their magic. Unlike one-shot flagship missions, the frequent-flying Discovery Program could accept risk of an occasional failure.[4]

The Discovery Program's great differentiator was its attitude toward money. Mission cost was capped at $150 million.

"The small scales and short lifetimes of these projects permit them to be terminated if they grow out of scope," NASA officials told Congress.[5] This contrasted sharply with NASA flagship missions, which, viewed as too big to fail, routinely ran up tabs hundreds of millions higher than promised.

Daniel Goldin, who took over as NASA's chief in 1992, became an outspoken champion of low-cost missions. Goldin's background building robotic spacecraft at the aerospace giant TRW contrasted with that of his astronaut predecessor, Admiral Richard Truly. Goldin latched on to the idea of "faster, better, cheaper" missions, to the point that the directive became known at NASA headquarters as "FBC."

"Let's see how many we can build that weigh hundreds not thousands of pounds; that use cutting-edge technology, not 10-year-old technology that plays it safe; that cost tens and hundreds of millions, not billions; and take months and years, not decades, to build and arrive at their destination," Goldin said.[6]

Goldin envisioned Discovery launches every month, with as many as one-third of them flaming out. This would be the price of moving faster and taking risks, Golden argued—a price much lower than "the scientific disaster or blow to national prestige that it is when you pile everything on one probe and launch it every ten years."[7]

In November 1992, about 250 scientists, administrators, and engineers met in San Juan Capistrano, California, for a week-long event aimed at rustling up missions for this new, faster-better-cheaper Discovery Program. They heard 73 presentations on ideas such as Venus landers; Mercury, Mars and Jupiter orbiters; moon studies; a cosmic dust collector; and eight comet missions.

Why such interest in comets? Scientists viewed comets as time capsules— remnants of an ancient, foggy disk of gas and dust that gravity and time shaped into planets and a new star. Eons of geological, atmospheric, hydrological and biological action had long since tainted the most accessible relic of that haze, Earth.

Billions of comets congregate in lazy orbits so far from the sun that their ingredients have remained in deep-freeze since the dawn of the solar system. When gravity from a planet or a star pushes a comet into an orbit visiting the sun, the comet's surface ices bake away, leaving a dark blanket of dust. But underneath that dust, comet experts believed, would be a storehouse of frozen compounds, as pure as they were when the comet formed with the rest of the solar system 4.6 billion years ago.

Scientists suspected that comets played a direct role in the evolution of life, which needed water and simple carbon-based compounds. Such chemicals were presumed to be scarce in the hot inner regions of the young solar system where the rocky planets Mercury, Venus, Earth and Mars formed. Perhaps comets served as delivery vehicles for the water and organic compounds now pulsing through rivers and our veins. Less disputed was the idea that distant relatives of comets plunged into the molten mass of our congealing protoplanet, their hydrocarbons and billions of tons of water to steam out over millennia of volcanism and then rain down and fill Earth's oceans.

This was theory, supported by little hard evidence. Scientists had believed for a half century that comets were a mishmash of ice and dust, or "dirty snowballs." But comets' true consistency remained a mystery, and an ice ball would tell a different creation story than a billion tons of something fluffy like freshly fallen snow.

THE DISCOVERY PROGRAM proposals included two ideas aiming to penetrate a comet's blackened husk to access its innards. One involved dropping an oversized arrowhead into the comet, with instruments inside sampling the material below the surface. A second, proposed by Marcia Neugebauer of the Jet Propulsion Laboratory, was titled simply "A Comet Impact Mission."

"An impactor is detached prior to encounter and impacts [the comet] just preceding flyby," Neugebauer wrote. "The impactor provides kinetic impact energy to produce a large crater and ejecta which are observed by trailing spacecraft and remotely from earth."[8] In other words, the spacecraft smashes part of itself into a comet while the other part watches the fireworks.

Neugebauer, one of America's pioneering female space scientists, had experience with comet missions. In the late 1980s, she had been a project scientist on the Comet Rendezvous Asteroid Flyby mission. CRAF was to track a comet for three years and hurl an instrument-laden spear into it. That mission, a companion to the Cassini Saturn orbiter, had been scrapped just months earlier, a victim of NASA budget cuts. This had been a blow to Ball Aerospace, which had been tapped to build the spacecraft for JPL.

NASA reviewers whittled the 73 Discovery Program proposals down to 11. Each of those won $100,000 to flesh out the ideas further. "These missions represent a bold new way of doing business at NASA," Goldin declared in February 1993. "By accepting a greater level of risk, we can deliver high-return missions that are cost-effective, quicker from concept to launch, and responsive to the present budget climate. They promise to revolutionize the way we carry out planetary science in the next century."[9]

Neugebauer's Comet Impact Mission, viewed as more risky and less scientifically compelling, didn't make the cut. But her idea would live on.

Michael Belton, an astronomer with the Kitt Peak National Observatory in Tucson, Arizona, had been a co-investigator on Neugebauer's comet proposal. He had independently proposed a Discovery Program idea called Small Missions to Asteroids and Comets, or SMACS. As to the purpose of SMACS—four small spacecraft that would zip past various celestial rocks and snowballs—Belton explained: "Exploration, discovery, and creative scientific research are the keys to new knowledge, and if we wish to know our origins and our destiny, whether we are unique or commonplace, and how nature governs our lives, we have no alternative but to explore the sun's system of planets, satellites, comets, and asteroids to discover their secrets and understand the processes that make them what they are."[10]

Belton won one of the 11 NASA grants. A team of crack comet scientists (among them Joe Veverka of Cornell University, Ken Klaasen of the Jet Propulsion Laboratory, and Mike A'Hearn of the University of Maryland) joined him in the SMACS scientific effort. But scientists don't build spacecraft. for that, Belton called an old friend.

ALAN DELAMERE, a senior engineer at Ball Aerospace, was a rainmaker, cultivating relationships with space scientists and landing contracts for the hardware of their dreams. He did as much business in hallways during space-conference coffee breaks and over dinners as he did from his Boulder office. Delamere had first met Belton in 1978 through a mutual colleague. At the Beckham Grill, an English-style tavern in Old Town Pasadena, California, Belton and Delamere had eaten bread pudding and talked about the comet Halley. Belton had asked whether Ball might design a camera for a joint U.S.-European mission to Halley in 1986. Americans were to build the primary spacecraft visiting Halley and another comet; the Europeans would contribute a Halley probe to piggyback and release for a close-up flyby.

As originally envisioned, the probe did not have a camera. Belton was arguing with the Europeans for a means of taking pictures of the most famous non-planet in the solar system, if only to help scientists figure out the probe's location in relation to the comet. At minimum, such context would enrich the data streaming in from the spacecraft's other instruments.

HMC 68 Image Composite
Comet Halley 14th March 1986

Halley's comet, as viewed by the Giotto spacecraft in 1986

An American budgetary wrangle killed the International Comet Mission. But Europe was unwilling to let Halley's first visit since 1910 pass without at least swinging by to say hello. They converted their probe into a self-sufficient spacecraft named Giotto. Delamere designed the Halley Multicolor Camera for the probe and led the German team building the camera. On March 14, 1986, their product became the first human creation to lift the veil of a comet's coma—a shroud of dust and gas thousands of times larger than the hunk of dirty ice feeding it—and see the nucleus beneath. It capture Halley in eerie profile, like a portrait of a poltergeist. Moments later, as if to avenge the interruption of 4.6 billion years of solitude, comet particles tore the camera's protective baffle from its mounts.[11]

The Beckham Grill had been a fitting meeting place for Belton and Delamere. Both men were born in England in the mid-1930s. Belton grew up in Gainsborough in Lincolnshire; Delamere, in Poynton, Cheshire. Both had served in the Royal Air Force. Both had built successful space careers across the Atlantic, though via different avenues.

Belton, the son of a news agent, graduated with a degree in astronomy from St. Andrews University in Scotland. Told that America was the place for astronomy, he came to the University of California at Berkeley in 1959. His PhD dissertation, published five years later, was on the effect comet tails have on interplanetary space. He participated in several missions, including the Viking program. Belton had a particular interest in calculating the rotation of comets, based on subtle differences in brightness as observed from Earth.

Delamere had graduated with an engineering degree from the University of Liverpool. He parlayed his experience flying as a 17-year-old volunteer reservist in the Royal Air Force into a job testing Blue Steel missiles—designed to carry British thermonuclear weapons to the Kremlin—as they guided themselves harmlessly into Cardigan Bay and, later, the Australian desert.

An avid outdoorsman and mountaineer, Delamere had his sights set on British Columbia. He made it as far as Hamilton, Ontario, where he took a job with Canadian Westinghouse. He came across Ball Brothers Research when he happened onto a Ball booth at a gathering in Detroit. In 1967, he arrived in Boulder for what he intended to be a five-year stay. He bought an old house on Mapleton Hill and never left.

Although his job was business development, Delamere was more idea man than a salesman. Good scientists have much in common with successful entrepreneurs. Both concoct schemes that seem crazy but turn out to be brilliant. However, while entrepreneurs generate ideas to make themselves money, space scientists come up with ideas that cost others money.

Delamere made his living seeking out—and being sought out by—modern-day John Lindsays. He had a mop of graying hair parted to the side, a wiry five-foot-eight frame, and a passion for Birkenstock sandals that burned through the winter. Delamere was charming, incisive and radically creative. He listened to scientists ramble on about their dreams, sifting their streams of consciousness for golden ideas. Then he would ask himself: "Christ, how the hell do we support him to make that happen?"

"Sometimes you regard the person as a lunatic for even thinking that something like that would be credible," Delamere said. "But we were very polite, and the net result was we went and thought about the problem from his vantage point and said, 'Hey, your idea can be made to work under the following conditions.' And we'd come up with solutions. And the scientists really appreciated it. They were not giving us a specification. They were giving us a problem to solve. And that's the difference between the way we worked and the way the other aerospace companies worked."

Delamere pushed his big ideas through the Ball system, coaxing managers to spend money. He worked the system like an experienced lobbyist. His enthusiasm earned him a reputation as a "wild child" and "loose cannon." A colleague described Delamere as having "the spark"—the sort of energy it took to do great things.[12]

Delamere understood space science to be a zero-sum game. He aimed to win, assessing the habits of NASA and the strengths and weaknesses of competing proposals, then refining Ball's response. He saw NASA's new Discovery Program as tailored to Ball's blend of science-instrument and spacecraft capabilities. Delamere had watched small teams with modest budgets solve problems with the Relay Mirror Experiment and Solar Mesosphere Explorer satellites and felt Ball could thrive under NASA's new program. Discovery could bring Ball something akin to a line of Orbiting Solar Observatories, the sort of repeat business the company had long sought but rarely landed. Behind Delamere's charge, Ball backed 16 of the 73 San Juan Capistrano proposals. After the meeting, Delamere told his bosses: "Our strong showing has convinced the key people at NASA HQ that Ball can be a significant player in low-cost planetary programs."[13]

Planetary programs—missions breaking Earth's gravitational grip, traversing deep space, and visiting the heavenly bodies beyond—represented a new frontier for Ball Aerospace. It had been 30 years since the jar company's unlikely leap into orbit. Its spacecraft had been stuck there ever since. Deep space represented not only commercial opportunity, but also the next big step in the evolution of Ed Ball's Boulder adventure.

MIKE BELTON'S SMALL MISSIONS to Asteroids and Comets was never to fly. In 1995, NASA selected the moon-orbiting Lunar Prospector and Stardust, a comet sample-return mission. (The inaugural Discovery missions, Near Earth Asteroid Rendezvous and Mars Pathfinder, had been grandfathered into the program.) Stardust would whisk past the comet Wild 2 in 2004, extending a sort of tennis racket in which the world's lightest solid, Aerogel, would grab comet particles. Two years after that, a Stardust capsule would land the dust samples in the Utah desert.

Belton was far from through with comets, though. Among other activities, he was chairman of a panel considering U.S. science instruments for a proposed European Space Agency comet mission—the ambitious Rosetta comet lander. The spacecraft would loiter near a comet for months, settle on its surface, and dig a meter down to sample the material. True, Rosetta was an exciting mission. But Belton believed the comet's treasures were buried deeper, beneath perhaps 10 meters of sun-scorched shell, and that only an impact could penetrate the crust. He hatched an idea indebted to Marcia Neugebauer's Comet Impact Mission.

One of the world's great cratering theorists worked across the street from Belton's office. Jay Melosh, a bearded, husky man, made his living at the University of Arizona building computer models of impact cratering. He had written the seminal book on the subject.[14] Belton paid Melosh a visit. Suppose a 500-kilogram projectile moving at 25,000 miles per hour struck a comet, Belton asked. What kind of hole would it make?

The answer depended on how hard or fluffy the comet was—something the impact itself was supposed to sort out. Melosh applied some standard equations and a few assumptions and came up with an answer: about 100 meters wide, he said, and maybe 30 meters deep.

"Hot damn. We're gonna do it!" Belton exclaimed.

Belton called a meeting at the Jet Propulsion Laboratory in Pasadena. Present were Melosh, members of the former SMACS science team, Delamere, and Rich Reinert, a Ball spacecraft designer. They had lunch. They talked about a comet-impact mission. Belton sketched out some rough ideas. By the time they left the room, they had named their next enterprise: Deep Impact.

Chapter 10

Misfire

Astronomers watch the universe passively. Deep Impact would be *active* astronomy, an attack on a celestial object. But what comet? Where? When? And, most importantly for Alan Delamere and Ball Aerospace, how?

The scientist-volunteers on Mike Belton's team knew why: to pierce through the comet's rind into the fruit within. The findings, they believed, would add hard data to theories about how our solar system had formed.

They chose the comet Phaethon. It had been one of Belton's proposed SMACS targets.

Phaethon was one of the 350,000 astronomical objects the Ball-built Infrared Astronomical Satellite telescope discovered in 1983. The three-mile-diameter, black-velvet body was thought to be an asteroid in a comet-like orbit until it was linked to the Geminid meteor shower.[1] Meteor showers generally happen when the Earth plows through the grainy wake of a comet. Phaethon was probably a comet masquerading as an asteroid.

Phaethon seemed dead. It lacked a typical comet's tail. Most comets shedded up to a meter of surface ice each time they passed near the sun. The lost material formed their tails. Phaethon passed closer to the sun than any other comet, about half the distance of Mercury's orbit (hence its name, after the son of Helios, the Greek sun god). It probably had been baked to the point that ices near the surface were long gone.

Confirming Phaethon's status as comet or asteroid was a fringe benefit of the mission, the scientists figured. Its deadness assuaged Delamere's fears of ballistic dust assaulting the spacecraft, as Halley had done to his Giotto camera. In addition, the software Delamere envisioned guiding the impactor would target the "center of brightness." Halley's bright halo, or coma, pushed the center of brightness well off the nucleus, as would happen with any active comet. Phaethon would be easier to target. So dead was good, the team concluded.

Bill Blume, a Jet Propulsion Laboratory orbital-mechanics specialist on the Deep Impact team, worked on a mission plan to hit the comet and then gather and send home enough data to make the whole effort worthwhile. The plan described an action sequence befitting a science-fiction movie. On March 13, 2002, with Phaethon racing across the darkness of space 132 million miles away, one of Deep Impact's two modules would smash into it at a relative speed of 74,000 mph.

The surviving spacecraft would carry three instruments. One would provide close-up views of the crater to help sort out the comet's density and constitution. A second, smaller telescope would take in the big picture. Inside each would also be a spectrometer looking at infrared light. With sunlight bathing the cloud of post-impact debris, infrared would reveal Phaethon's chemical secrets. A third telescope, riding on the back of the spacecraft, would photograph Phaethon after the flyby.

To come up with a spacecraft design, Delamere enlisted the part-time efforts of several colleagues—electrical, mechanical, and optical engineers. They worked the details of the proposal for a year, with internal business-development funds paying for the Ball engineers' time. Ball Aerospace and its competitors spent small fortunes this way. Ball invested roughly $250,000 in the Deep Impact proposal, and the company had dozens of proposal teams going at any given moment, often with different scientists competing for the same NASA prize.

Among those pouring their energies into Deep Impact was Rich Reinert, the Ball Aerospace spacecraft designer. Reinert had a reputation for both speed and creativity in turning scientists' vague notions into viable spacecraft concepts.

Reinert had graduated from MIT with a degree in aeronautical and astro-nautical engineering in 1968 and had been in the aerospace business since. He was graying, mustached and had a stentorian voice. He spoke as if addressing a roomful of engineers.

At Hughes Aircraft in 1972, Reinert designed a window for the Pioneer Venus lander. Sapphire had been used for other windows, but the gem would partially blind the infrared instrument looking through this particular window. Diamond, he figured, could survive Venus's hostility—pressures 100 times higher than on Earth's surface, temperatures of 900 degrees and a corrosive atmosphere—while still allowing infrared light through. Jewelers in Rhode Island carved a disk the size of two stacked pennies out of a 200-carat, $2.5 million, walnut-sized gem imported from Amsterdam. It was the color of a dark beer bottle, Reinert remembered. The U.S. government reimbursed Hughes tens of thousands of dollars in excise taxes once the spacecraft launched. "As far as the State Department was concerned, we had re-exported it to Venus," Reinert said.

Later, while at Boeing, Reinert worked on a study to explore space-based nuclear-waste disposal, dreaming up five-foot-diameter containers capable of surviving a launch-pad explosion. Computer simulations showed that radioactive waste in an orbit halfway between Earth and Venus would stay put for 50 million years. "At that point, you can build a house out of the stuff and live on it," Reinert said. The Department of Energy took the proposal seriously, he remembered, but found blasting nuclear waste into space carried about the same risk as burying it and would cost twice as much.

Reinert had visited Boulder in the 1970s. Ball Aerospace built the infrared instrument that peered through his diamond window. He joined the company in 1984 to be a part of what he saw as a self-selected group of refugees from big companies. "What Ball really does—our thing—is doing something no one has done before," Reinert explained.

Working for Ball was different in other ways, too. "There's never any question whether Boeing is going to be there in five years, but there's a question of whether you'll still be employed by Boeing at that time," Reinert said. "At Ball, it's a fundamentally different proposition. There's three jobs for every person, so you're never bored. And then the other thing about Ball: while you knew you'd still be employed there in five years, it was never clear whether Ball Aerospace would still be there in five years. So it's a whole different outlook."

THEODOR VON KÁRMÁN, the aerodynamicist and co-founder of the Jet Propulsion Laboratory, once said, "Scientists discover the world that exists; engineers create the world that never was."

Engineers love this quote. But it ignores the reality that most engineering design—most design in general—is about subtle improvements to some long-established, successful template. Design an automobile, a hamburger, or a newspaper, and there are clear rules—four wheels and brakes, bread surrounding meat, headlines above text. In contrast, the engineers concocting a comet-bashing spacecraft found even the basic grammar open to debate. Such a blank slate could paralyze an engineer. Reinert felt liberated.

There were constraints, of course. A Boeing Delta II rocket could only lift so much and fit a finite bulk in its nose cone, or fairing. Being a Discovery Program proposal, Deep Impact had to cost less than $183 million, up from $150 million in 1992, but still small potatoes for planetary science missions.

Reinert also knew he could stand on the shoulders of giants. Before Ball's first orbiter could fly in 1962, Pete Bartoe had to invent dual-spin stabilization, and Marion Fulk had to concoct a lubricant to keep Bartoe's gyroscopic inspiration from seizing up. Reinert had more than 40 years of spacecraft design to inform his notion of Ball's first deep-space bird. Modern attitude control engineers had mastered three-axis stabilization and used it on most spacecraft (the OSOs had been stable in just two dimensions), applying various techniques pioneered in the late 1960s.

Spacecraft designers profited from other technical advances. Solar arrays of the mid-1990s could supply three times the energy per square foot as in the days of the Orbiting Solar Observatory, giving designers more freedom to support power-hungry instruments or squeeze more capable spacecraft into the noses of rockets. Batteries held triple the energy per pound. Computers, which had reshaped American society and economy in the intervening decades, had similarly revolutionized spacecraft. The first OSO responded to ten commands; modern spacecraft acted on thousands.

In designing spacecraft, Reinert stayed tuned to the workings of his subconscious mind. "What you do is you look back upon all the spacecraft you've designed before and you study the requirements of the current mission and these basic conceptual ideas," Reinert said. "And somehow when you initialize the problem like that in your head, it all works in the background and usually the next morning in the shower you'll have a vision of exactly what the thing should look like."

He sketched ideas and handed the sketches to computer-aided designers. Reinert sometimes put together three-dimensional mockups made of cardboard, masking tape, tin foil, paper, and string. He would take that to the Ball model shop, where the fabricators built better mockups out of wood, metal and plastic.

For Deep Impact, Reinert borrowed ideas from a satellite Ball was building for the Navy, called Geosat Follow-On, or GFO. GFO, to be launched in 1996, was an ambitious spacecraft, capable of measuring sea level to within an inch while flying 480 miles over the water at 17,000 mph. Thanks to a carbon-composite structure, GFO weighed only about 900 pounds fully fueled. Its body was shorter than a refrigerator and topped with a flat white antenna reminiscent of something a studio photographer might use to soften the light. A solar panel about the size of a ping-pong table was bolted on via struts.

About seven hours before crossing paths with Phaethon, the GFO-like Deep Impact mother ship would release an 1,100-pound, camera-laden aluminum can into the target's path. Lacking navigation or propulsion systems, this "impactor" spacecraft would strike within 3 yards of its intended aim, Delamere calculated.

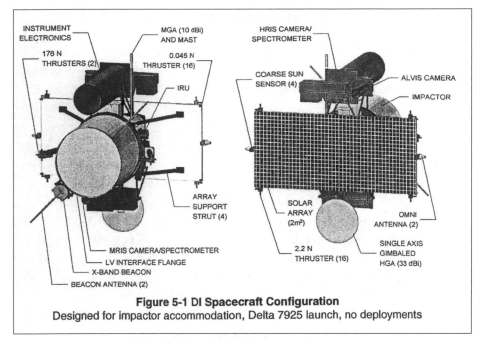

Figure 5-1 DI Spacecraft Configuration
Designed for impactor accommodation, Delta 7925 launch, no deployments

The Deep Impact spacecraft, as envisioned in the 1996 proposal

There was precedent. In July 1995, while en route to its orbit around Jupiter, NASA's Galileo spacecraft released a probe incapable of adjusting its course. It drifted for five months before disappearing into the storms of Jupiter.

Delivering a force equivalent to 64 tons of dynamite, Deep Impact's craft would blast a crater about 30 yards deep and 170 yards in diameter, the team estimated. But it was conceivable that the blow might demolish Phaethon like a piñata. "It would be a major achievement for Deep Impact should splitting occur," scientists noted.[2]

The surviving spacecraft, called the "flyby," would flee the collision course as soon as it released its cargo, and have about seven minutes to watch the aftermath before Phaethon screamed past.

The flyby would spend the next 18 days sending home images and spectrographic data about the chemical makeup emanating from the blast. It would all cost $158.2 million, they figured—well below the Discovery Program cost cap.

On December 5, 1996, Belton, Delamere and their Deep Impact team sent a document no thicker than a holiday catalog to NASA, titled "Deep Impact: An Exploration of the Surface and Deep Sub-Surface Structure of a Cometary

Nucleus." It was one of 34 proposals various teams sent to NASA, including three others backed by Ball Aerospace.

In April 1997, NASA narrowed the list to five, each to receive $350,000 for a four-month, in-depth study. Those proposals included missions to Mercury and the Martian moons Phobos and Deimos, a solar-wind study, a mission to Venus, and a comet mission. The comet mission was Contour, which would fly past at least three comets, photograph their nuclei, and analyze the dust flowing from them. Deep Impact lost.

NASA's selection committee was interested in Deep Impact, but worried about the ability of a rudderless can to strike the comet in a bright spot. It could miss entirely, but even a shadowed hit would be a problem. Deep Impact's cameras and chemical-analyzing spectrometers relied on sunlight. NASA also questioned Phaethon's suitability as a target: what if it was an asteroid and not a comet? Why send a comet-hunting spacecraft to an asteroid? NASA ranked proposals from one to four. One meant the proposal was worthy. Four meant, "Forget it, fellas. Don't ever propose this again," as Delamere put it.

Deep Impact got a four.

The scientists were disappointed, but even with more launches encouraged by the Discovery Program, the demand for ideas always outpaced the supply of hardware and rockets. Proposing space missions was like pitching Hollywood screenplays. Most would never be made.

Some members of the Deep Impact team had found success elsewhere. Deep Impact science team member Joe Veverka of Cornell was the lead scientist for the Contour comet-flyby proposal, and four other scientists from the failed Deep Impact bid had signed on as co-investigators.

Ball was out a quarter-million dollars. Delamere, a Deep Impact co-investigator, had other work to do. He was on the finalist Venus mission's science team, which Ball had supported at his urging. But Deep Impact wasn't over yet. Delamere and others still wanted to crack open a comet to see what was inside.

Chapter 11

Let There Be Life

NASA panned Deep Impact. Belton and Delamere could have moved on and just let it die.

But as the 1990s Internet boom took root, as Hong Kong switched from British to Chinese control, as Lunar Prospector—NASA's first competitively selected Discovery Program mission—took orbit around the moon, Belton and Delamere resuscitated their comet mission.

They took their dead-comet, dead-impactor proposal and infused it with life. The new spacecraft would have a live impactor. The live impactor would steer itself into a live comet.

A real death changed things for the team proposing Deep Impact. Mike Belton's wife Helyn succumbed to cancer in 1997. Belton, bereaved and exhausted, felt that continuing as lead scientist was more than he could bear. He wanted a smaller role on the science team, focused on establishing the comet's rotation. Mike A'Hearn took over.

Mike A'Hearn

The University of Maryland astronomer was among the world's foremost authorities on comets. He had studied the mysterious bodies since the late 1960s. He had devised the standard set of filters that observatories around the world used to watch Halley's comet in 1986. A big man with thinning white hair and a thickening white beard, he looked like an off-duty Santa Claus in Birkenstocks and rumpled shirts unbuttoned to the sternum. He had no experience with space hardware, much less with space hardware comprising a major NASA program.

Belton and A'Hearn knew comets well enough to have a short list of candidates for targeting. But so little was known about comets—only Halley had ever been photographed up close, thanks to Belton's initiative and Delamere's camera—they weren't going to be picky.

Don Yeomans, a Jet Propulsion Laboratory scientist who had joined the Deep Impact team, considered potential targets. Yeomans had written a book on the history and folklore of comet observations. His day job involved leading the Jet Propulsion Laboratory's Near Earth Object Program. The group kept tabs on thousands of asteroids and comets to identify those on a collision course with Earth.

Yeomans tracked comets as well as anyone in the world. In 1982, he guided astronomers to the first sighting of Halley's Comet in 71 years, when it was still more than 900 million miles from Earth.[1] For this new project, Yeomans chose the object Tempel 1.

Comet orbits are tricky. Planets shove them around. Cometary ices thaw and jet off into space, which alter a comet's rotation and path. If Halley's comet had been a rock, it would have arrived four days earlier in 1986.[2] Even with a much shorter orbit of five-and-a-half years, Tempel 1's spewing of ices would delay its arrival by several hours. Ignoring the effect of outgassing would cause Deep Impact to miss its target by more than the distance separating Earth and the moon.

ERNST WILHELM LEBERECHT TEMPEL discovered Tempel 1 in 1867. It was among 13 comets and five minor planets found by the self-educated, itinerant German astronomer. Don Yeomans and the Jet Propulsion Laboratory were familiar with Tempel 1. It was the proposed target for an in-progress JPL comet mission called Space Technology 4/Champollion, which was to rendezvous with Tempel 1 in 2006.

A moderately active comet, Tempel 1 passes near the sun 14 times more frequently than Halley, which flies beyond Neptune's orbit before turning back around.

Tempel 1 was a short-period comet, its orbital arc swinging only roughly as far as Jupiter. Also known as Jupiter-class comets, Tempel 1 and its ilk are believed to come from the Kuiper Belt, a donut of billions of comets and larger frozen bodies, including the dwarf planet Pluto. A gravitational push from a giant outer planet occasionally sends Kuiper Belt comets hurtling toward the sun. Jupiter's enormous gravity keeps them hemmed in until their surfaces bake so deeply that, like Phaethon, the comets appear to be asteroids.

Long-period comets are also worth noting. Hale-Bopp, the "Great Comet of 1997," passed near the sun so rarely that it glowed 1,000 times brighter than Halley at a similar distance.[3] Since its last visit, Hale-Bopp had flown ten times father than Halley. If the souls of the 39 members of the Heaven's Gate cult hitched a ride with Hale-Bopp after their infamous 1997 mass suicide, they won't return for 2,500 years.

Comet Hyakutake, a surprise in 1996, traveled 20 times farther yet. Long-period comets like Hale-Bopp and Hyakutake can orbit as far as 50,000 times the Earth-sun distance, with round-trips lasting millions of years. Unlike Tempel 1 and other short-period comets, which travel around the sun in roughly the same plane as the planets, long-period comets come from all corners. Their home, theorists say, is the Oort cloud, a spherical haze of icy bodies enveloping the solar system like hands around a firefly.

Oort-cloud objects, scientist believe, formed near Jupiter, Saturn, Neptune and Uranus, but were cast off to the solar system's boondocks, slingshot by the powerful gravity of these gas giants. Oort cloud comets are so far away that gravity from passing stars brushes them into interstellar space or toward the sun, the theory goes.

TEMPEL 1 WOULD REACH ITS PERIHELION, or closest point to the sun, in early July 2005. Tempel 1 whirled around the sun canted about 5 degrees from the imaginary plane occupied by the planets. The comet spent half its time "above" the solar system and half its time "below" it. Tempel 1 would cross the plane at perihelion. This was decisive, the mission designers knew.

Moving out of Earth's orbital plane takes an enormous amount of energy. Deep Impact would have to meet Tempel 1 at the key intersection of perihelion regardless of when it came. JPL's Bill Blume, while reworking the Deep Impact mission design, ran the numbers. He settled upon July 4, 2005, as the day Deep Impact must meet Tempel 1. A mission whose most notable feature was celestial fireworks would, by pure coincidence, have its comet encounter on Independence Day.

The changes to the proposed spacecraft centered on the impactor. Rather than an oversized aluminum can, the new spacecraft would look like a rusty oil drum. The science team realized a vaporizing aluminum impactor would blind the telescope's infrared instruments to certain coveted spectrographic information about the comet's makeup. The impactor would be made of copper instead.

"Copper's in the column in the periodic table with gold and silver and it's cheaper," A'Hearn explained. "And that column says it doesn't react with the water in the comet."

But the copper was only skin deep. The impactor's aimlessness had helped sink the 1996 proposal. This time the spacecraft could calculate the comet's location and, equipped with thrusters, steer into Tempel 1. A "cold-gas" propulsion system would use puffs of nitrogen, just like on the first Orbiting Solar Observatory. The smart impactor also needed a computer brain. It needed ears and a voice, so it got a radio to communicate with the mother ship. And it got eyes—or at least a single eye. Alan Delamere, serving again as Ball's proposal manager, and Rich Reinert, again on spacecraft design, added to the stout impactor an exact copy of the flyby spacecraft's slender, six-foot-long high-resolution telescope. It would be better for comet close-ups, and it gave the small spacecraft the look of a motorcycle with a sidecar.

Blume's mission design, although similar to his Phaethon plan, had important rewrites. The spacecraft would move into the path of Tempel 1 as the comet raced by at 66,000 mph. The impactor, moving in the same general direction but much more slowly, would vaporize against the comet at a relative speed of 23,000 mph. That's fast, certainly—more than ten times the speed of a rifle shot. But it was a fraction of the 74,000 mph relative speed with which Phaethon would have mowed down the impactor. That difference in velocity, combined with intelligence and agility embodied in the new-and-improved impactor, allowed the flyby and impactor spacecraft to separate a full 24 hours before the main event rather than the 7.2 hours of the previous proposal. The flyby would burn less fuel scrambling away from the onrushing comet and have twice as long—14 minutes—to watch the impact's aftermath before turning a well-armored cheek to Tempel 1's ballistic dust as the spacecraft passed as close as 310 miles.

Dead comets shed no dust, so the Phaethon spacecraft had needed no protection. To face off against Tempel 1, the new Deep Impact carried an 8-foot, 106-pound Kevlar-and-aluminum disc, a deep-space play on Captain America's mighty shield. It would protect the flyby spacecraft from rice-grain particles packing the punch of assault-rifle shots.

Reinert and Delamere expunged the look-back imager, deciding instead to turn the entire spacecraft so the two telescopes could capture photos of the comet after it raced by. The solar panels grew from two square meters to three in order to power the impactor's new systems until they were released a day before the encounter.

The flyby and the impactor spacecraft would target the same spot on Tempel 1, communicate for a time, then, during the final minutes before impact, settle on the same bright spot and adjust their relative trajectories.

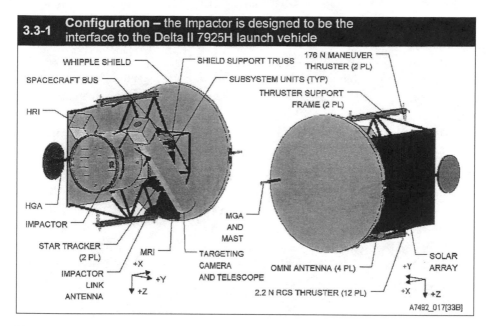

3.3-1 Configuration – the Impactor is designed to be the interface to the Delta II 7925H launch vehicle

The Deep Impact spacecraft, as envisioned in the 1998 proposal

The impactor would strike the comet; the flyby, still as far from the comet as Athens, Greece, is from New York, would swoop in and aim its scopes so precisely that it could record the layers of the crater as Tempel 1 zipped past. The chain of events would unwind 83 million miles from Earth, nearly as far away as Earth is from the sun. It takes more than seven minutes for light—and therefore radio communications—to make the trip, which translates into a 15-minute delay from sending an order to getting word back from the spacecraft. In the mission's most critical moments, the spacecraft would have to steer themselves, via software the mission team would have to develop.

The whole Deep Impact package would cost NASA and the American taxpayer $233 million, they estimated, well below the Discovery Program limit, now $299 million. It was a bargain for two spacecraft and deeper knowledge of our origins, the team believed. On June 29, 1998, Mike A'Hearn sent the proposal a few miles down the road to NASA Headquarters on the banks of Interstate 395 in Washington, D.C. "Deep Impact: Exploring the Interior of a Comet," its cover read. Spanning 46 pages plus appendices, it was among 26 proposals responding to NASA's call. It also, by chance, had the same name as a big-budget Hollywood movie released the previous month. The movie featured a seven-mile-wide comet with its eye on Earth. Rather than strike their target with a copper spacecraft, though, the movie heroes drilled nuclear warheads into it.

As the seasons changed, A'Hearn and his colleagues went about their business. They got word in early November that, this time, Deep Impact would get a closer look. The comet hunter was among five proposals atop NASA's shortlist. Deep Impact's competitors included a mission to gather samples of Martian moons by firing slugs into their surfaces and flying through the debris cloud; a Jupiter orbiter to study the giant gas planet's interior; a Mercury orbiter; and a Venus orbiter focusing on the cloudy planet's middle atmosphere. Each would get $375,000 for a four-month concept study. The winner, or winners, of the next round would receive millions.

"Deciding which one or two of these exciting finalists will be fully developed will be a very difficult choice," said Ed Weiler, the NASA executive who would ultimately narrow the list.[4]

DEEP IMPACT WAS STILL a big gamble, but now one with better odds. Delamere's proposal had been the equivalent of dangling a worm in the water. Now, a huge fish was on the line. Jerry Chodil was the man to determine how to get it in the boat.

Chodil, a Ball Aerospace vice president, was born in a working-class neighborhood of Chicago to a machinist father and a homemaker mother who worked part-time as a cashier at the A&P. He earned a masters degree in physics from the University of Chicago in 1963 and, interested in nuclear physics, headed to Lawrence Livermore National Laboratory in Livermore, California.

There he worked with a group charged with launching sounding rockets from the Hawaiian island Kauai. Instruments on the rockets measured X-rays, gamma rays, and other potent radiation emanating from high-altitude nuclear explosions, like the Starfish Prime blast that had scorched the first Orbiting Solar Observatory. The Americans, Soviets and British had signed the first atmospheric test ban treaty months before Chodil's arrival. So the group practiced their art by looking for X-ray-emitting stars with the same rockets and detectors. Instruments on the Livermore group's Aerobees discovered neutron stars and pulsars one after another.

In 1968, Chodil returned to Chicago, taking a job with Zenith and earning several patents on a research project involving flat-panel gas-discharge displays—now known as plasma TVs. Convinced it would take thirty years for the technology to reach the market and that the electronics-savvy Japanese, and not Zenith, would be the ones selling it, he started looking around. Time would prove him right on both counts.

Ball hired him in 1976. Delamere was Chodil's first boss at Ball. He asked Chodil to look into using charge-coupled devices, or CCDs, as the digital "film" of a space-based camera. Bulky TV cameras had delivered every electronic image ever sent home from an unmanned spacecraft. CCDs, based on semiconductors that convert light energy, or photons, to electrons a computer circuit can make sense of, would change that. Delamere's Giotto camera captured its images of Halley on them ten years hence. In the 1990s, CCDs revolutionized terrestrial photography, displacing film and endowing billions of mobile phones with sight.

Chodil rose through Ball's ranks and by 1998 led all of Ball's NASA-related efforts. When word arrived that Deep Impact was a finalist for the next Discovery round, Chodil decided to "pour the coals on this thing."

"Spend whatever you need to spend and I'll worry about the money—you guys just go and win this goddamn thing," he told his team.

Although Delamere and Reinert continued to play central roles, Chodil brought in Ron Young to lead a concept-study team of more than a dozen full-time engineers. They would flesh out Delamere's and Reinert's rough-sketch proposal with the aim of showing NASA that their comet mission could be made to work, and for a price in the ballpark of the proposal's original estimate.

Young, 60, had led several concept-study proposals in his 30-year career and had a knack for winning them. He was a slow-talking native Iowan approaching retirement, and so avid a fly fisherman that he co-owned a Boulder fly shop. Fly fishing and space-mission proposals had something in common. To land a fish, you have to think like one. When pitching space missions, Young tried to get into NASA's head. He also considered the competition's most likely gambits and highlighted things he suspected they wouldn't. And showmanship was a must. "You've got to come up with some grabbers," Young said. "Some pizzazz. Some excitement."

Deep Impact promised pizzazz as few proposals did. It pledged to smash into a comet on a day which, through a stroke of luck, happened to be America's pyrotechnic holiday. Its scientific aims reached beyond chemistry and geology and into questions of theological gravity: Where did we come from? What are we made of?

Young knew grabbers when he saw them. The initial proposal had stuck to the facts, focusing on the science and the engineering to deliver it. The concept study would be a more aggressive sell. Under the heading "Benefits to NASA" atop the concept study's executive summary, science still won top billing: "Views deep into the interior of a comet nucleus will allow answers to fundamental questions, while other comet missions only scratch the surface." But the second bullet point was "huge public appeal."

"The explosion of a 500-kg impactor into Comet Tempel 1, with near-real-time image return, will naturally attract the public and school children as no event since Mars Pathfinder," it read, referring to the hugely popular 1997 Mars lander mission. Pathfinder, the second Discovery effort, had bounced onto the red planet's rocky surface in a cocoon of airbags, unfurled itself like a daisy, and released the two-foot-long Sojourner rover, humanity's first interplanetary motor vehicle.

The Ball team and its scientific partners knew NASA scored such missions on a point system, and they scrounged for every possible tally. They talked about building with proven hardware, reusing software, developing new technologies, working with women-owned and minority businesses, and promoting the mission to the public. Desperate to score points for technology transfer, the team said they would use experimental fault-protection software from the Deep Space 1 mission. They touted a "10/10 Experiment," which would allow outside scientists to compete for a 10-kilogram instrument using no more that 10 watts of power. They wrote of a tie-in with the Discovery Channel involving a small video camera. There would even be a Deep Impact theme song.

The concept study grew into a heavy tome combining fancy graphics with pages of spreadsheets squeezing text and numbers to the brink of illegibility. Young, Delamere and the proposal team estimated the cost of every piece of hardware and human task they could imagine, basing their prognostications on experience from previous missions and educated guesses. It was an exercise in detailed soothsaying, with the hope that all the overestimates would pair with underestimates and balance out in the end. "You'll get some areas where you haven't got a clue, really, because it's so different from anything you've ever done before," Delamere admitted.

Deep Impact's design saw some important changes. The circular ballistic shield gained 20 pounds and became rectangular. The massive telescope hanging astride the impactor was replaced by a copy of the smaller flyby telescope and dubbed the Impactor Targeting System, or ITS. It was small enough to live inside the copper barrel, making things more elegant visually, better balanced physically, and less prone to damage from random cometary assaults. The flyby's instruments, mounted at a 45-degree angle to the direction of the impactor's release in the proposal, were now boresighted straight down, such

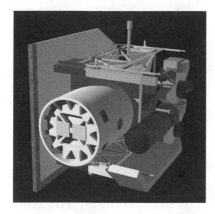

The Deep Impact spacecraft, as refined in the 1998 concept study

that the spacecraft could release its mate without having to reorient, a move inherently bad for accuracy.

It would all cost more: $273 million. NASA's tab would be slightly less, $269 million, with Ball agreeing to invest the balance in equipment needed to build the spacecraft and in software for a next-generation spacecraft operating system. JPL threw in $650,000 in software. The grand sum included 22 percent—$50 million—in a giant bucket of contingency, for all the surprises guaranteed to crop up as engineers fleshed out the design. Included in the appendices was a Letter of Endorsement from Chodil to Mike A'Hearn, assuring the mission's principal investigator that the budget "should cover any unforeseen problems that may arise during the development phases of the mission."

Privately, Chodil was less certain. "When you propose on something, you can't imagine how much stuff you really don't know yet. Enough details haven't been worked out yet, all the drawings haven't been generated, all the hardware and subcontract stuff isn't coming in yet. There are so many unknowns, it should scare the pants off of some people and it does," he said. "I was the guy who signed off on the proposals and I can't tell you how many times I'd sit in my office and they'd come in for a signature and I'd think to myself, Wow, this is going to be a tough one. But I also took the attitude that it's the nature of the beast. Either you're in this business or you're out of this business."

On March 26, 1999, A'Hearn sent NASA his team's concept study, which was predominately the Ball team's work. It looked like any other two-inch-thick, plastic-bound corporate report. A million dollars had been poured into it. Chodil had come up with more than $600,000 of Ball money to make sure Young had his pizzazz.

ED WEILER'S JOB was to make sure the $2 billion Americans gave each year to explore the cosmos was well spent. He had most recently guided the agency's science efforts with respect to the origins of life and the universe, which generally involved big space telescopes peering so far away the light itself was billions of years old. More importantly for Deep Impact's fortunes, he had spent

20 years as NASA's chief scientist on the most famous of all telescopes—"The People's Telescope"—the Hubble Space Telescope.

Weiler was an astrophysicist with the bearing of a corporate executive. Short, stocky and towheaded with bright blue eyes, he had a directness that was disarming or startling, depending on the situation. He maintained a youthful spark, with a quick-trigger laugh generally accompanied by the wide-grinned look of a mischievous child.

Ed Weiler

The son of a steelworker, he grew up in inner-city Chicago, earned his PhD at Northwestern in 1976, worked for the great Lyman Spitzer at Princeton, became a world authority on a very particular type of binary star, and then left the publish-or-perish world of academia in 1978 to fulfill a lifelong dream of working for NASA. A year later, the 30-year-old Weiler became the Hubble Space Telescope's chief scientist.

Hubble would become the greatest thing in astronomy since Galileo raised his telescope in 1609. The great space telescope helped nail down the universe's age (13.7 billion years), spotted planet-factory dust clouds around most young stars, multiplied by thousands the estimates of how many galaxies are in the universe, and provided key insights into the speeding-up of the universe's expansion and the dark energy causing it.

Such discoveries depended on Hubble's infrared and ultraviolet instruments detecting various bands of light our eyes can't see. But the pretty pictures of dazzling spiral galaxies, looming nebulae, and sparkling clusters made the telescope one of the most famous human implements of all time.

Astronomers loved Hubble, too, but it was widespread public affection that kept the space telescope near the top of NASA's priorities for more than a decade. Americans spent billions of dollars and shuttle astronauts risked their lives replacing gyroscopes and batteries, switching out old instruments for new ones, and otherwise propping up the aging giant.

Weiler was filling in as associate administrator for the Office of Space Science when NASA announced Deep Impact as a Discovery finalist on November 12, 1998, and he officially won the job four days later. He knew Ball Aerospace well—the company had built five instruments for his space telescope over the years, including Murk Bottema's vision-correcting inspiration. More importantly for Deep Impact, Weiler understood, as well as any scientist alive, what stokes the flame of public interest.

In a fifth-floor NASA Headquarters conference room, Weiler sat down with his budget director and the directors of the astrophysics, solar physics and planetary science divisions he oversaw. They gathered on a summer day in 1999 to give birth to $600 million in space-science missions. Thick concept studies were piled on the table. The Deep Impact team and its competitors had made passionate cases. *They* should send the Earth's next robotic missionaries to Jupiter, Venus or Mercury; *they* should build the first spacecraft ever to poke at the moons of Mars or to bonk heads with a comet. All had been classified Category 1, *crème de la crème*. According to the NASA rule book, Weiler and his team could have chosen all five if they had had the money. They didn't have the money.

The group quickly settled on the Mercury orbiter as the clear standout. The sun's nearest neighbor had remained unexplored since Mariner 10 whisked by it in 1975. The mission to the moons of Mars paled in comparison. The Venus proposal hung on longer, but ultimately failed to measure up to Inside Jupiter or Deep Impact.

With the American economy in Internet-bubble overdrive, NASA could afford two new Discovery missions. It could not afford three. Deep Impact, its detractors insisted, had less science to offer than a detailed study of Jupiter's innards. The Inside Jupiter mission was to orbit the giant planet and map out its internal structure using measurements of Jupiter's gravitational and magnetic fields. Comet missions, doubters argued, had stocked the Discovery manifest already, with Contour scheduled to photograph the bodies, and Stardust, chosen back in 1995, slated to collect particles from the tail of a comet and return them to Earth.

Others countered that the billion-dollar Galileo spacecraft was orbiting Jupiter at that very moment. While Jupiter's remaining secrets remained tantalizing, the planet was hardly unexplored. Besides, Contour would look at a couple of comets and Stardust would bring home spatterings from the cooked shell. There would be valuable science, of course. But only Deep Impact could get beneath the surface of orbiting bodies so foreign to science and so central to understanding the very formation of Jupiter and the rest of the solar system.

Weiler listened as the factions pressed their cases. Mission selections usually depended on consensus. This had the markings of a bold, slice-the-baby-in-half executive judgment. It was going to be Weiler's call, among the biggest in his brief tenure.

One of the other big decisions had been announced a few days earlier. Weiler had euthanized a mission whose costs had spun out of control—Space Technology 4/Champollion, the spacecraft that had brought comet Tempel

1 to Don Yeomans's and JPL's attention. Champollion was to orbit Tempel 1 and then send a lander to the comet's surface to perform shallow mining and chemical analysis. Its leaders had promised a low-cost mission for trying out such novel technologies as inflatable solar-panel booms and solar-electric propulsion. But Champollion had grown into a multinational cometary-science extravaganza, and costs ballooned accordingly. Canceling that one had been easy, once Weiler got past the NASA administrator and the Congressional delegations.

Eliminating Champollion had been fiscally prudent. Plus, it didn't hurt Weiler to have a cancellation under his belt. The worst thing an administrator could do was threaten to cancel missions and never follow through, Weiler believed. Henceforth, project managers would know that Weiler's threats were not empty ones.

So here was yet another visitor to Tempel 1. While he had no particular warmth for the frozen body, he knew a bargain when he saw one. Plus, Deep Impact had…*pizzazz.*

Yes, the science, the cost, the risk and the sheer believability of the proposal—can they really do this for a price in the same zip code as what they're quoting?—had to line up. But there was also the Hubble factor. Weiler was both a father and a child of Hubble. He had a feeling this *thing*, this crazy comet mission, was going to capture America's imagination, more so than anything anyone would send to study the innards of Jupiter or anything else.

NASA's connection to the American people relied to no small extent on journalists. Weiler, NASA's public face for Hubble, had been dealing with them for two decades. He had spoken into countless microphones and had stared into the cold eyes of cameras for crews from *60 Minutes, Nightline, Today, Good Morning America,* and many others. He knew what made the mainstream media tick.

"There are certain things that sell—sex, violence and anything with 'biggest' or 'smallest,'" Weiler said. "And this was an explosion. A cosmic explosion. It was going to be front-page news. And it's not just the explosion. It was primordial material. It's the stuff we're made from. It had all the nuggets of good science public affairs, and I think Hubble has demonstrated to NASA the value of good, quality public affairs and what it can do for the whole agency."

He won over enough colleagues at the table to ensure that his decision would not be viewed as executive fiat. There was no consensus—his planetary-science chief wanted a look inside the jumbo planet, period—but Weiler had won a majority.

On July 7, 1999, six days after Weiler and NASA quietly killed the Champollion mission to comet Tempel 1, NASA announced a new Discovery mission to the same destination.

Ball Aerospace would have had its first shot at deep space either way, as it turned out. The Jet Propulsion Laboratory had tapped the Boulder company to build the Inside Jupiter spacecraft that Deep Impact had narrowly edged out. Ball had lost; Ball had won. Such was life in the space business.

Deep Impact, risen from the dead, was a go. Forty years had passed since tiny Ball Brothers Research began work on its first sun-seeking orbiter. The company finally had its long shot into deep space. Now they just had to build what they said they would.

Chapter 12

Devils in Details

Legend had it that Ball engineers designed the five-story Tech Tower in the 1960s. The exterior looked like a Soviet apartment building. Inside, the men's and women's bathrooms alternated floors—except on the fourth floor, where a customer complaint had brought a ladies room in the 1990s. There once had been no women's bathrooms at all. The tower's designers had never imagined women would work in the labs. The structure afforded its occupants low ceilings and little sunlight, with the exception of an outside band of offices such as the one in which Lorna Hess-Frey knitted her brows.

At first glance, Hess-Frey violated all stereotypes of the mechanical engineer. She had long sandy-blonde curls and an unmistakable physical vigor. One imagined her outdoors, maybe on a horse against a snow-capped mountain backdrop, perhaps teaching her two young children how to ride. She had grown up on a 2,000-acre ranch near Durango—riding animals, sewing her own wardrobe, and driving balers.

The young rancher had been good at math and science, and when she was a junior in high school, an engineer from a California bubble-gum factory had come in for Career Day. He described the plant's ventilation system, a vast network of pipes coated in powdered sugar, and she was hooked.

Hess-Frey had just recognized a problem with the pipes for the proposed Deep Impact spacecraft. Rather than powdered sugar, the issue was toxic hydrazine. Hydrazine, N_2H_4, was a spacecraft propellant of choice. It produced thrust without fire. A chemical reaction stoked at a couple hundred degrees broke down and rearranged nitrogens and hydrogens into ammonia (NH_3) and then just nitrogen and hydrogen. Such cracking released energy, which propulsion engineers could direct so precisely as to aim a spacecraft at a speeding comet.

Hydrazine flowed from a propellant tank to thrusters in quarter-inch-diameter stainless-steel tubes. The Deep Impact concept study had tubes running everywhere, reaching out into space, under debris shields, even through the solar panels. Hess-Frey knew the hydrazine exhaust could settle on a solar panel or a telescope mirror and stick like glue. In addition, the tubes would have to be heated, because hydrazine freezes at about the same temperature as water. In space, frozen hydrazine would paralyze the thrusters.

Heaters for the tubes would need more power than the single solar panel could deliver without starving the rest of the spacecraft. The solar panel would barely fit inside the nose of a rocket as it was. Either the panels had to grow, or the hydrazine system had to be revamped. It would be a major redesign regardless.

There was more. The third stage of a Boeing Delta II rocket, the latest incarnation of the one that once carried the Orbiting Solar Observatory into space, spun at about 70 revolutions per minute. Deep Impact's weight would either balance on an axis within roughly a millimeter of its physical centerline or, once released from the whirling Delta third stage, begin with gentle wobbling and end with a $270 million addition to the great junkyard in the sky. It's what a washing machine on spin cycle would do if it had the wherewithal.

Hess-Frey looked at the solar panel on one side and the bulky instruments on the other, at the latticework of propulsion tubes and the 128-pound Kevlar-aluminum debris shield, and knew this contraption would never balance, at least not without an Olympian counterweight. It was so unbalanced, in fact, the spinning might rip it apart. It was an amazing oversight to have somehow slipped by everyone working on the concept study, not to mention NASA evaluators.

Indeed, the flyby spacecraft Rich Reinert and Alan Delamere had dreamed up could never fly. Certain systems looked fine—the instruments, for example, seemed to be in good shape, maybe even overdesigned—but Hess-Frey and others recognized that Deep Impact could look nothing like what was proposed.

There were no recriminations. Hess-Frey herself had done the concept study's impactor design, and it was way off the mark, too. With the news that NASA had picked Deep Impact, Jay Melosh, the cratering expert, wasted little time in modeling the copper barrel's behavior on impact. He found that, when colliding with Tempel 1 at 23,000 mph, the manhole-sized caps on either end of the impactor gobbled up gigajoules of precious kinetic energy as they crunched into each other and vaporized. The cylindrical sheath would plunge into the comet like a straw into a sno-cone, disgorging relatively little in the way of pristine materials. Hess-Frey, without the benefit of such insights, had proposed a dud of an impactor.

Indeed, some of the best engineering minds at Ball Aerospace, with much consultation with the vaunted Jet Propulsion Laboratory, had, on closer inspection, proposed a mule and sold it to NASA as a thoroughbred.

It happened all the time. A proposal was about plausibility. Reinert had designed all variety of spacecraft over the years, each as tenable as the next. None saw the nose of a rocket. Sometimes he came closer than others. The fact was, Reinert and Delamere and others involved in the proposal had no time to actually engineer anything. Their goal was to suspend NASA's disbelief, to show beyond a reasonable doubt their design was innocent of transgressing the laws of physics and the rules of accounting. The conceptual designer's job was to put something together that *looked* like it might work.

NASA, which had been buying spacecraft for 40 years, understood this. And Ed Weiler was no fool. Changes were fine in his view. Deep Impact just had to stay below his fiscal line in the sand, or he would kill the mission as swiftly as he had ended Champollion.

While Hess-Frey pondered the overall shape of the craft, power engineers considered electrical needs; computer engineers calculated throughput and storage; software engineers ran through options for code; telecommunications engineers mulled antennas and amplifiers; optical engineers focused on mirrors and prisms and lenses; and attitude control engineers considered spacecraft pointing. Dozens moved their offices to the Tech Tower, where they soon consumed the top three floors. Deep Impact was by far the biggest thing Ball Aerospace had going in early 2000. Then there were teams at JPL working on Deep Impact's interface with the Deep Space Network, as well as the fault-protection software and the autonomous navigation software system that was to guide the impactor to the comet.

Hess-Frey's team of 12 mechanical designers was at the center of it all. Deep Impact needed form before other teams could nail down the spacecraft's various functions. She found herself in meeting after meeting, informal chat after informal chat. The Tech Tower's whiteboards became riddled with sketches of wild-looking machines. She handed off design ideas to underlings who used computers to plug in the assumed components on spacecraft based on triangles, boxes, pentagons, hexagons, octagons, chevrons and other forms.

At the same time, at the urging of Melosh and other scientists, the impactor changed, too. The scientists wanted a copper sphere, which would blast a really nice hole, they said. But copper weighs about four times as much as aluminum and is, metallurgically speaking, butter-soft. Nobody builds copper spacecraft, much less a spherical one upon which 1,000 pounds of flyby spacecraft was to ride during a rocket launch. Hess-Frey and her colleagues drew the line. No sphere.

Ball's long tradition of working with scientists prevailed. There was a "cratering workshop" where engineers and scientists hashed out the finer points of smashing a comet. The engineers listened to the scientists' ideas and tried to accommodate them, going as far as concocting such designs as a "14-sided cuboctahedron," which would have been roughly as easy to build as it was to pronounce. The targeting instrument would stick out of the cuboctahedron's flank like a toe through a worn sock.

They finally settled upon a shape approximating a giant rivet, a meter wide and a meter tall, its curved surface to be made of layered plates of hollowed-out copper, the back end holding the instrument and electronics. That design, a compromise, would deliver a scientifically gratifying impact despite a boxy behind an engineer could love, one capable of housing instruments and supporting the flyby spacecraft's weight.

Jeremy Stober came up with the rivet idea. He was a big six foot three with

The final impactor spacecraft design

glasses and a short goatee, and was not yet thirty. He had studied aerospace engineering at the University of Colorado, coming out in 1992 in the depths of an aerospace bust. He found a job at a Denver environmental-engineering firm, where he did such things as model amorphous plumes of groundwater pollution.

Stober did modeling at Ball, too. He arrived in 1994 and worked proposals early on. He became a self-trained expert in three-dimensional computer-aided design software and turned Rich Reinert's pencil-and-paper sketches into classy renderings other engineers could work with and present to the outside world.

On Deep Impact, Stober worked for Hess-Frey. He rearranged parts on various possible visions of Deep Impact's dual spacecraft. With clicks and rolls of a mouse, he shifted space hardware around like the features of Mr. Potato Head dolls. An instrument here, an antenna there, thrusters hither, solar panels yon. Achieving spin balance and supporting heatable propellant tubes were only the beginning of the limits he faced. The flyby spacecraft had to first firmly grip and then release the redesigned rivet impactor, for one. The solar panels, which had grown to the size of two ping-pong tables, somehow had to fit into a nine-and-a-half-foot-diameter rocket fairing. The instruments, with their heat-sensitive infrared detectors, needed constant shade. There had to be ballistic shielding, but it could not be the size of a sliding glass door anymore. Antennas had to point toward Earth, and the instruments, Delamere insisted, were to aim in the direction of the impactor's release. Structural simplicity—the elegance in which Pete Bartoe had so firmly believed—and ease of maintenance also came into play.

The final Deep Impact design

Months of iteration yielded more than a hundred designs, then winnowed to twelve, then to three, then to one known simply as Concept 7B. Not a triangular prism, nor a box, but rather a pentagonal column would define Deep Impact's aluminum-honeycomb exoskeleton. Inside, the electronics and computers and fuel tanks and propellant lines would have a bit more protection from comet wash and be easier to keep warm. Five sides offered enough room on the inside and a good angle for folding the spacecraft's wing-like solar panels such that they could squeeze into the rocket nose. Plus, should the solar panels refuse to snap open—40 years of space heritage and Vac Kote notwithstanding, moving parts remained the bane of spaceflight—they would still gather sufficient light.

Delamere could only shake his head. He had intended to build Deep Impact so inexpensively, using so much off-the-shelf hardware that the mission could actually give money back to NASA at the end of the project. He knew spacecraft evolved as details emerged with hard study, but this…this was a much more complicated creation than he had ever imagined. The heritage from Ball's earlier satellite, the Geosat Follow-On, had been thrown to the wind. The spin-balancing oversight had been unfortunate, but it was certainly no showstopper. Dick Woolley, though retired to a Denver condominium, could be called in. Dick Woolley could spin-balance anything!

And the instruments. They pointed not straight down toward the impactor, as Delamere had wanted, but rather at a 45-degree tilt. Hess-Frey and Stober told him this was a necessary evil if the instruments were to fit in the rocket nose and somehow stay in the solar panels' frigid shade.

But had not Delamere insisted, over and over, "We can't take our eye off the ball"? The comet was maybe three miles across. Until a couple of days before impact, even its much more massive halo would be a speck in the eye of the biggest telescope deep space had ever seen. Tempel 1 would come in like a fastball. You need to watch and calculate and recalculate your course until the last possible second…and *then* release the impactor as gingerly as possible. That way, even a dumb impactor would probably hit the comet. This Concept

7B spacecraft would watch and calculate and recalculate just fine. But then it would pull its eyes from the target a day before impact, to lean 45 degrees and drop its copper load. It would take its eye off the ball. If the impactor for some reason failed to steer itself properly, they would have less than a day to fix it. A miss was all but guaranteed. Plus, the whole release exercise would complicate the mission's choreography and add expensive hours to the jobs of the JPL and Ball engineers charged with guiding and aiming the spacecraft. It could be made to work, but at what cost? This was a Discovery mission. The biggest risks weren't technical. They had to do with schedule and budget. Screw those up and you'll never get off the ground.

BILL FRAZIER HAD GREAT RESPECT for Delamere. Ball owed Deep Impact and many other programs to the man. But Frazier had seen problems from the moment he laid eyes on Deep Impact's initial concept. He had come off a classified Ball program to take over as Deep Impact's lead systems engineer. It was his job to anticipate everything that could possibly go wrong and make sure it didn't happen.

Frazier, 43, was the son of an electrical engineer and a mathematician. His parents had met while working at Lawrence Livermore National Laboratory. Frazier had been on the University of Colorado ski team as an undergraduate but later gravitated to triathlons and other endurance sports more in keeping with a work ethic that frequently had him hitting 40 hours by Wednesday afternoon. He had joined Ball in 1983 with a master's degree—perhaps the only real relaxing he ever did had been a semester off in those years, skiing and waiting tables in Aspen—and earned his PhD in aerospace engineering in two years while working part time at Ball.

He had spent six years as the lead system engineer on the Geostat Follow-On spacecraft on which Deep Impact's proposal had been based. He knew GFO as a super-light, low-Earth-orbit bird made of carbon composite, a difficult material to shape into a bicycle frame much less a spacecraft. Ball had built precisely one carbon-composite spacecraft, Frazier said, and would probably never attempt it again. Either way, applying such a design to something supposed to lug an 1,100-pound copper bullet millions of miles didn't make sense. This flyby spacecraft was way too light to wrestle with the impactor. It also lacked the electrical and computational horsepower for the job.

Now they were talking about an entirely different spacecraft design—a new baseline—that borrowed much less from previous missions. Keeping

Deep Impact anywhere near the promised $270 million would be a huge challenge, Frazier knew.

Some solutions were straightforward. Frazier became convinced scientists had settled on an 1,100-pound impactor—an even 500 kilograms—because it sounded good. True, the heavier the impactor, the bigger the crater. But the laws of physics promised that dropping the impactor from 500 kilograms to 350 kilograms—a 30-percent cut—would only shrink the crater about 10 percent. Plus, the impactor's weight loss was enough to downgrade Deep Impact's Boeing rocket from a Delta II "heavy" to a standard model. It would save $6.6 million. Mike A'Hearn and his science team acquiesced.

The instruments, Frazier and others saw, were full of extra electronics and complicated redundancies that begged to be pared back. Rather than build a $1.4 million "scan mirror" to sweep across the post-impact debris plume, why not just rotate the spacecraft? The big and small telescopes both had $500,000 spectrometers. Do we really need the one on the small instrument? Using radio amplifiers left over from JPL's Cassini Saturn orbiter would save more than $1 million, as would downgrading the antennas the flyby and impactor spacecraft would use to talk to each other. A planned medium-gain antenna was dropped. The flyby spacecraft's 20 thrusters were pared back to eight. The cut would make the job of engineers steering the spacecraft harder, but would clear the thicket of vulnerable propulsion tubes and make Deep Impact easier to build.

Some of the wilder ideas in the concept study disappeared. The proposed 10/10 payload melted away, as did a deal with Discovery Communications, which would have involved a video camera to film the impactor's separation and also record Tempel 1 from up close.[1]

But for every dollar saved, the project seemed to be spending two more. The spacecraft redesign had set the project back nearly $1 million. The solar panels having to pop open meant another half-million dollars in latches, explosive bolts and testing. The more the JPL and Ball teams thought about it, the more they realized their gyroscopes lacked the sensitivity needed to point at the comet, and they hadn't even considered accelerometers, which were needed to understand precisely how much a given thruster pulse changed the spacecraft's trajectory. Add $1.7 million.

The biggest change, one that would cost Deep Impact tens of millions of dollars, stemmed from a single comment uttered at a design review. NASA projects ran two gauntlets early in their lives, a preliminary design review—a four-day marathon for Deep Impact—and, about a year later, a critical design review to assess the final engineering plans. In the space business, their abbreviations, PDR and CDR, were as familiar as BMW or IBM. Get past PDR and you started placing orders with companies making major hardware like spacecraft computers and solar panels. By CDR, the detailed design was all but done, and spacecraft construction was moving ahead full-bore. Stumble at either review, and the mission was over.

Both reviews involved at least two dozen NASA consultants visiting Ball for weeks. They were the space-business equivalent of expert witnesses, often semi-retired graybeards with long experience building spacecraft, and they were all looking for blemishes on the proposed comet hunter. The consultants shared an uncanny ability to spot problems simply by the way engineers described them. *That was vague. What exactly do you mean by that?* At one such meeting, someone wondered aloud what would happen if comet pebbles happened to kill the flyby spacecraft just after the cosmic explosion but before it had a chance to transmit its data trove to Earth.

What, indeed? The mission plan was for the impactor and the flyby to take thousands of images and spectra on approach, store them in a hard drive or flash memory, duck and cover during the comet flyby, and then, when the all-clear sounded, spend a few leisurely days sending back a mountain of scientific data. The science team's comet-dust models assured everyone that the chance of a sniper granule strike was at most five percent. But the risk was real. Look what Halley had done to Delamere's camera. What a shame it would be to spend all that time and effort to hit the comet and take all those lovely pictures—only to have them stuck in a dead spacecraft forevermore.

The collective response came to be called the "Live for the Moment" strategy. Prior to the collision, as the impactor's camera and the flyby spacecraft's pair of cameras and spectrometer snapped away, the flyby would gather and bundle its own data plus the impactor's data and, rather than simply storing it, send it back to Earth, frenetically filming the impact's aftermath and stuffing its data pipes, capturing, saving, packaging, and sending as much information home as it could before turning to brace itself for the cometary squall. That way, even should the comet take out the flyby, the mission's most vital data would make it home.

But all that capturing, saving, packaging and sending—while navigating and thrusting and monitoring such things as power and temperature—would take more computing power than Deep Impact had. The spacecraft computer was to have been based on a microprocessor called the RAD6000. The radiation-hardened chip had first flown on the Mars Pathfinder spacecraft and was battle-tested. But the Live for the Moment strategy meant Deep Impact would need something better. The team settled on something called the RAD750. It was based on a PowerPC chip of the sort sold in Apple Macintosh systems several years earlier. On Earth it was already a dinosaur. But BAE Systems, an aerospace contractor, had gone to the trouble of toughening up the processors so solar and cosmic rays wouldn't kill them—a certainty without special conditioning. Deep Impact's RAD750 would be the fastest processor in space, a full ten times faster than the RAD6000. But it had never flown.

Deep Impact, then, was now a first-of-a-kind mission involving two spacecraft, torrents of data, and a new brain. Or rather, three new brains—two on the flyby (there had to be a backup) and one on the impactor. The flyby's two brains would be heavily cross-strapped in case something went wrong. More than simply having a backup you fired up like a diesel generator during a power outage, a cross-strapped system ran more or less in parallel. Both computers went through the same motions, letting each other know what they were doing all the while, although only the primary actually had its hands at the controls. If the A system froze up or otherwise shied from its duties, the B computer would step in like a stunt double. Developing and testing such a Siamese-twin system in a creature as complex as Deep Impact would be an enormous challenge, Frazier knew.

Navigation was another concern. Deep Impact's success depended on not just plowing into the comet, but doing so in a well-lit spot visible to the flyby's instruments. The autonomous navigation system would consider images from the impactor and flyby, calculate an average "center of brightness," and go for it. But the center of brightness would differ depending on the comet's shape and texture (round? dumbbell-shaped? smooth? pockmarked?), about which little was understood.

Here, the Deep Impact team had exactly one source of information: a composite photo made from 68 Halley images Delamere's Giotto camera had taken in 1986. The team knew Tempel 1 would be charcoal black and that it would probably be cratered from various collisions during its eons in orbit. But whether Tempel 1 was a packed slush ball or like cigarette ash or an orbiting Cocoa Puff, nobody knew. Don Yeomans, the JPL comet-orbit expert, once sketched a speckled, dented ball with a hole straight through it. Through the hole he had drawn an arrow. Such an outcome wasn't likely, but it *was* possible.

On top of it all, there was the deep-space factor. Ball Aerospace may have been building low-Earth orbiters since the first OSO, but this was different. Frazier, ultimately responsible for the spacecraft's form and function, saw Ball as a place with a lot of smart people who had no idea what they were getting

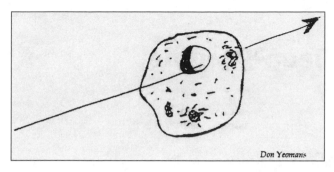

An unlikely, but possible, Deep Impact mission scenario

themselves into. Ball was a deep-space infant, crawling about under the watchful eye of JPL, the Boulder company's partner in Deep Impact.

Late one night, working again after he had put his kids to bed, Frazier found himself distracted from the task at hand. This thing was *so* complicated, he thought to himself—dual spacecraft, new computers, new software, new instruments, autonomous navigation—and it all had to work right, the first time. It never worked right the first time. Improving the mission's odds would take an enormous amount of testing absent in either the budget or the schedule of this faster-better-cheaper mission. That testing would depend on software and hardware unlike anything Ball had ever built.

And what if it didn't work? What if Ball, after clawing its way to the big leagues and finally standing at the plate, just whiffed? With, ultimately, his team—him—responsible?

Chapter 13

Deep Trouble

A project manager's days are locked in chains of meetings in which discussions focus on interpreting the contents of spreadsheets. Project managers organize, track, evaluate, decide; then, based on the numbers, their experience, and their intuition, they try to motivate people. Then they update the spreadsheets and prepare presentations summarizing such things as "status."

It's not sexy. Most project managers, even the best of them, whose products win world renown, remain anonymous. Who was project manager on the Hoover Dam? The Boeing 747? The Statue of Liberty?

Brian Muirhead rode into Pasadena on his BMW motorcycle. In 1977, he strode onto the storied, buttoned-up campus of JPL in sandals. To a place where the influence of the freewheeling 1970s had only just begun to jostle crew-cut tradition, Muirhead brought a worn leather bag, a beard, a ponytail and some serious attitude.[1]

The Jet Propulsion Laboratory prided itself on its ability to cherry-pick the top students from the country's most prestigious schools. Muirhead, prior to graduating from the University of New Mexico, had repaired Mercedes Benzes in Albuquerque and motorcycles in Greece. But he was supremely confident, and JPL soon realized they had landed a rare talent. Fifteen years later, ponytail shorn but bearded yet, Muirhead became flight system manager on Mars Pathfinder, similar to the role Bill Frazier played on Deep Impact.

Dan Goldin, the NASA chief, had challenged JPL, the nation's primary purveyor of gigabuck spacecraft, to land a machine on Mars, complete with rover, for less than $270 million and in three years, tops. The last time JPL or anybody else landed on Mars, with twin Vikings of the 1970s, it had cost $1.5 billion, which equated to about twice that in modern money, and it had taken six years. Goldin was poking JPL in the chest. *You think you can do faster, better, cheaper? Prove it.*

JPL delivered a spectacular success. On July 4, 1997, Pathfinder landed on Mars and set its 25-pound Sojourner rover free. Sojourner examined red rocks mission control named "Yogi," "Barnacle Bill," and "Prince Charming." The national media and the public lapped it up. NASA basked in a restored glow. Muirhead, energetic, outspoken, charismatic and vital to the triumph, appeared on about every national news show, won the 1997 *Design News* "Engineer of the Year" award (the magazine titled the accompanying profile "Mr. Mission Impossible"), and even wrote a business book called *High Velocity Leadership* on doing faster-better-cheaper the Mars Pathfinder way.

NASA's Ed Weiler reminded Muirhead of his mortal status in mid-1999—Muirhead had been project manager of JPL's cancelled Space Technology 4/Champollion mission. The setback was brief, with Muirhead soon reappearing atop the organization chart for NASA's new mission to the comet Tempel 1.

If Muirhead was a whirlwind, his employer was a weather system. JPL had more than 5,000 employees on its 177-acre campus and was recognized as the world's top interplanetary space organization. Like Ball Aerospace's, its origins were humble. In 1936, a couple of young researchers from the Guggenheim Aeronautical Laboratory, California Institute of Technology, or GALCIT, fired handmade rocket engines in a parched creek bed of the Arroyo Seco, which would have been just up the road from the Rose Bowl if there had been a road. The group struggled for a couple of years; then the U.S. military got mildly interested; then V-2 rockets rained onto Allied Europe; then the money started pouring in. In 1944 they came up with the name "Jet Propulsion Laboratory," though flying turbines were never the point. The term "jet" avoided the lingering crackpot stigma of rocketry.[2]

JPL built Explorer 1, America's response to Sputnik, and never looked back. The lab sent Rangers to the moon; Mariners to Venus, Mars and Mercury; Voyagers to zoom past Jupiter, Saturn, Uranus and Neptune; and Galileo and Cassini to hang around Jupiter and Saturn and their many moons. JPL's space telescopes spied the heavens; its Earth observers watched over land and seas. Its Deep Space Network, with 34-meter and 70-meter-diameter ears in California, Spain and Australia, could hear the whispers of spacecraft billions of miles away and boom instructions right back to them. JPL was unique in the world, enjoying billion-dollar annual budgets and representing "a substantial national investment in money and brainpower," as the historian Peter Westwick put it.[3]

JPL was part of NASA—officially speaking, a NASA Center, paid for by the federal government, like Goddard or Marshall or Johnson—and apart from it, managed by the California Institute of Technology. So JPL never quite fit in. Whereas the wild-child Goddard Space Flight Center of John

Lindsay's day matured into a semi-cooperative NASA adult, JPL remained a sort of rebellious teenager in the agency that fed it. It was government, but it wasn't government. It was in the space business, but it wasn't a business. "The *raison d'etre* for government-owned labs, such as JPL, was to undertake long-range R&D projects that industry could not do or would not do," Westwick observed. "JPL trained its eye on the cutting edge and beyond, not the bottom line for shareholders."[4]

Another historian simply called JPL "a university that develops spacecraft instead of students."[5]

For the Deep Impact mission, JPL was NASA's feet on the street. The University of Maryland had its principal investigator and Ball had its project manager, but in NASA's eyes the buck stopped with JPL—more specifically, with Brian Muirhead.

Muirhead's counterpart at Ball Aerospace was Ron Young, who had led the Deep Impact concept-study proposal team. Young was an erstwhile valedictorian who spiced discussions about hydrazine propulsion systems with the word "garsh," as in, "Garsh, it's almost lunchtime." He liked his work; he loved to travel and to fish the Bighorn up in Montana, where he and his wife had land and a trailer.

Ball had a habit of assigning winning proposal managers to build the actual spacecraft, a tradition that reached back to Fred Dolder and the first Orbiting Solar Observatory. It tended to keep proposal managers from making outlandish promises. Young had worked on 11 spacecraft in his three decades at Ball, many from proposal to finale. Before that, he was part of Boeing's Saturn V rocket team. He had a track record of success. He assembled Ball's Deep Impact team, many of whom he had worked with back on the concept study. Together, they were embarking on a mission that seemed to him to be just…different.

Deep space was new to Ball, but what got Young was the uncertainty: nobody knew if the comet was light like cotton candy or hard like cement, cratered or smooth, or shaped like an apple, pear, banana, or donut. The greatest comet minds in the world were on the science team, and they just shrugged their shoulders. *Garsh, we don't know.* And then you've got a customer—Young and his colleagues tended to call JPL "The Customer," a term embodying emotional neutrality, reverence or derision depending on the moment—with a very different way of doing business.

Muirhead took over in 1999. On one of his first Deep Impact trips to Boulder, Muirhead had dinner at the home of Carl Buck, one of the people who had seen Pathfinder through.

Born in the town of Fowler in southeastern Colorado, Buck had come to Boulder as a boy when his newspaperman father got a job working at the *Daily Camera*. His parents split up. His dad went back to Fowler and Buck stayed with his mom. When he was 15, Buck rode his bike miles out to the Boulder airport after his Sunday paper route. He helped pilots work on their planes. One pilot was a big, gravelly type who drove a Triumph, flew a De Havilland Tiger Moth, and wore yellow aviator glasses. The plaque on his plane read: "O.E. 'Pete' Bartoe."

The Ball Brothers Research president took the rudderless young man under his wing. Buck ended up owning an airplane before he had a car—a Bartoe-designed Skyote biplane he bought and built himself. He trained as a draftsman at vocational school and took classes at a community college. Bartoe put a good word in for Buck when he decided to transfer into the University of Colorado engineering school, where Bartoe was an esteemed alumnus.

Buck joined JPL in 1983 and worked in Pasadena as a mechanical engineer and manager for 25 years, many of them under Muirhead. But Buck's wife had taken a job in Boulder, and he was ready to come home. Rather than leave JPL, he worked on the Ball-JPL Inside Jupiter proposal, which Deep Impact had edged by a nose. Muirhead saw Buck as a perfect man to be JPL's on-site presence for the Deep Impact project. Buck was one of those people who seemed like an old friend from the moment you met him. For someone in the Ball-JPL crossfire, it was a qualification second only to competence.

After dinner, Buck and Muirhead moved to the family room and talked about the project ahead. Deep Impact had money and weight to spare, they agreed. Compared to Mars Pathfinder, this thing was going to be easy.

REALITY POUNCED like a tiger. The technical challenges were extreme at any price. But making things worse, in Muirhead's estimation, was that the initial proposal had been underbid by tens of millions of dollars. It had been way too lean, involving way too many assumptions carrying way too much risk of failure. Even the concept study, which had added $36 million to NASA's proposed Deep Impact tab, left the program a good $20 million shy of what it should have been, Muirhead felt. The wholesale spacecraft redesigns were proving that out. Just over a year into the mission, Muirhead's spreadsheets told him the project could end up *$32 million* over budget. Deep Impact was running far too hot.

From the beginning, Muirhead saw a lot of JPL in his Boulder counterparts—bright, talented people, energy, *esprit de corps*. "My Pathfinder perspective said the only way we were going to do this was by a very unique, highly interdependent partnership with Ball. And I was excited about that," Muirhead said.

He convened two team-building events in the mountains above Boulder. A consultant came in and talked about levels of "teamness." He drew a graph on which the line formed something like a check mark. Groups with well-defined master/servant relationships performed at the lowlier left side; high-performing teams climbed to the Promised Land on the right. In the valley between languished those who were supposed to be teams but didn't act like them.

Deep Impact appeared to be stuck in the trough. As similar as Ball and JPL engineers may have been, as dedicated as they all were to the shared goal of success in space, a fundamental difference in heritage was tainting the relationship.

JPL was a nonprofit. Ball was a business.

JPL enjoyed enormous resources. Ball Aerospace was on a tight corporate leash, operating in a low-margin, risky realm of commerce. Adding to the challenge was Ball Aerospace executives' accountability to Ball Corp. shareholders, many of whom had bought the stock for consistent returns on Ball's steady aluminum-can and plastic-bottle businesses, which comprised 85 percent of corporate revenues. Sales of beer and soda were little influenced by political winds or the whims of government officials, whose decisions were the difference between huge wins and bitter disappointments in east Boulder.

But the big issue was the two organizations' very different perceptions of risk. Engineers are passionately risk-averse, for which all who have crossed a highway bridge can be thankful. But they are not infinitely so, if only because cutting risk—by making things stronger, more flexible, less error-prone—costs money.

JPL had specialized in multi-billion-dollar, must-not-fail missions for decades. Such an environment demanded an intimate understanding of a mission's risk—the sum total of the foibles lurking in each nut, bolt, wire, line of code, component, and subsystem; in the interaction between subsystems; and across the spacecraft as a whole as it launched into and lived out its days in the ultimate beyond.

JPL had little taste for risk. Its engineers considered how the best system they could possibly build would perform, then looked at how much they could actually spend and whittled back the design. If you went straight for a lesser system, how could you *really* understand how much risk you were accepting?

The great rub, of course, was that risk, like love, was ultimately incalculable and, unlike cost or schedule, subjective. Risk was in the eye of the beholder.

Alan Delamere had a hearty appetite for risk and put together a proposal reflecting as much. So did NASA chief Dan Goldin, at least in word.

Another factor was that JPL knew to its organizational marrow that failed deep-space missions won painful notoriety, bringing with them embarrassing press conferences and withering criticism.

Failure was no more welcome at Ball. It was bad for morale and worse for business, or at least would be—the company told all who would listen that it had never suffered a mission-ending failure. But Ball, with 2,500 employees in an industry where some competitors had 50 times as many, was also lean—"great starting lineup, not much of a bench," as one JPL manager put it—and targeted risk with arrows just large enough to bring it down. They felt that JPL went after risk with bazookas, which burned engineering hours and money and led to similar solutions in the end anyway.

The Boulder company wanted in its heart to be a servant to JPL's master. In Ball's view, one made money in aerospace by understanding the requirements, figuring out what it will cost, and then asking The Customer for a good deal more. But with Deep Impact, the combination of JPL's predilections, Ball's lack of deep-space experience, the uncertainty surrounding the comet, and the sheer magnitude of the effort—"This was not a little instrument job. This was a major spacecraft," Muirhead said—quashed any realistic hope of such a simple arrangement.

Young sensed something askew with the relationship. Dealing with The Customer as an equal "teammate" seemed... *weird*. But Young liked Muirhead, whom he considered "a top-notch guy, technically and personally." They became friends.

Such was the stage upon which a Deep Impact team that was not really a team set out to build two spacecraft destined for a cometary target nobody quite understood. Then machines began smashing into Mars.

THE FIRST WAS THE MARS CLIMATE ORBITER, a weather satellite for the Red Planet. Lockheed Martin built the spacecraft in Colorado. As with Deep Impact, JPL managed the project and was in charge of navigating the spacecraft. A Lockheed engineer had written a bit of software code that gave the spacecraft's thrusters instructions in English units—pounds—rather than the metric Newtons the JPL's navigation team expected. When the thrusters fired, they did so with more than four times the resolve that they should have. JPL navigators noted the quirk before launch and during the cruise, but never followed up with Lockheed Martin.[6] The orbiter swung around Mars on

September 23, 1999, ending its life as a $125-million shooting star as it burned up in the Red Planet's atmosphere. When word of the cause came out a week later, the chairman of the U.S. House Committee on Science and Technology released a two-word statement: "I'm speechless."[7]

The Mars Polar Lander followed on December 3. A hyperactive sensor and an injudicious software patch caused the lander's engine to shut off while the spacecraft was still 130 feet in the air. It touched down at close to highway speed. A trip of 470 million miles ended in a cloud of reddish dust.

Bill Frazier, the lead systems engineer, had been gleaning what he could from old friends down the road at Lockheed Martin. Lockheed, besides perhaps JPL itself, was the world's premier builder of spacecraft destined for deep space. Yet they suffered two crushing failures in a row. What did that say about Ball's chances?

Both Mars missions had been faster, better, cheaper. More cross-checking would have prevented both disasters. Goldin's FBC had put NASA itself in the cross-hairs once again. "America's space program *lacks* a big picture," wrote the CBS News anchor Dan Rather. "'Faster, cheaper, better' suggests utilitarian efforts conducted in the low background of the national consciousness—space as commerce, or space as bureaucratic exercise."[8]

NASA commissioned accident reviews, which hammered the agency. Goldin told an audience at JPL, "In my effort to empower people, I pushed too hard and in doing so stretched the system too thin. It wasn't intentional, and it wasn't malicious. I believed in the vision, but it may have made some failure inevitable."[9]

Ed Weiler, the NASA space science chief, found himself testifying before Congress about missions he had inherited. Enough, he decided. "The primary lesson was: Don't be penny wise and pound foolish. If you've got a three or four or five hundred-million-dollar mission, spend an extra two or three or eight million dollars to make sure you've got the proper oversight. Make sure the navigators are talking to the designers. Make sure that you're using metric units throughout," he said. Otherwise, he continued, "It can cost you a mint. It can cause a failure on Mars."

Deep Impact, bid as a faster, better, cheaper mission with all the risks inherent in such corner-cutting, was now something else. Muirhead calculated a need for more than $8 million on "Directives to Increase Mission Success" alone—$1 million for cross-checking, $3 million for independent verification, and $300,000 to "improve communication with contactor." Deep Impact shared with the Mars Climate Orbiter the worrying trait of having the spacecraft team in Colorado and the navigation team in California.

The directives' hidden costs were much higher yet. Alan Delamere's frustration had mounted as his influence waned on the mission he had brought to Ball and JPL. His concept-study spacecraft had been rubbished in favor of a much more extravagant creature. Muirhead, with Young's complicity, was pushing up the price tag dangerously, in Delamere's view. Delamere noted a fundamental change in the project's tenor. Deep Impact had been proposed as a mission that traded low cost for high risk, and now, he said, "We had the wise men telling us we must not fail."

"So the 'must not fail' empowered everybody, all the way down to the power-supply people, to say, 'I've got to make sure it works, not to stay within costs,'" he said. "So there were two cost factors. One is the extra work you have to meet the extra pushups that are being done. And the second was the change in the mindset of the people that *it must not fail.* So both those things caused this huge cost escalation."

RON YOUNG LEFT DEEP IMPACT in early 2001 and retired a few months later. Having spent 40 years in the business and now atop a volatile mission, he felt it wiser to climb down and spend more time casting for trout. Ball management also sensed the program was slipping. They called in John Marriott.

Marriott, 46, was no more dashing a foil to Muirhead than Young had been. To Muirhead's surprise, his new counterpart had never managed a spacecraft project. Marriott had been an instrument guy and, before that, a quality inspector. He had started at Ball without a college degree, having been an Air Force technician. He had gone to night school in Denver and earned his diploma in business administration, an unconventional choice in an environment ruled by engineers. He was a pool shark, could see five moves ahead when he played chess, owned a take-n-bake pizza joint on the side, and had a knack for picking stocks. He was a different animal on the Boulder campus, a place where Ed Ball and David Stacey had set an enduring tone of collegiality and politeness.

Behind the glasses and the graying temples, Marriott's face was unlined and even boyish. But his bullish aura borrowed something from that of a mob boss. "Come into my office," he would say. Then the deep, gravelly voice, uttering words so laden with intent they seemed to thud onto the desk: "I need you to do this," or "We need to make a change," or "You guys are screwing up," he might growl. Marriott viewed project management as a battle on many fronts. He had come to the same conclusion as Niccolo Machiavelli, the leadership philosopher who taught his prince that while being loved is good, being feared is better.

Marriott worked extreme hours, holding meetings and poring over detailed schedules to try and get his head around what was happening. He crammed on the ins and outs of spacecraft building, too, because he knew he couldn't manage something he didn't understand. He saw an unrealistic schedule and ballooning costs, the products of underestimates, redesigns and the overhead added after the Mars failures. This mission was in trouble. JPL was trying to build the best spacecraft money could buy, it appeared to Marriott. "And our guys were saying 'Yeah, good idea.' And they *were* good ideas, but unfortunately there was no money to do that," Marriott said.

He found TBDs—To Be Determineds—growing like weeds in the mission-requirements document, which serves as a space mission's Organic Act, its constitution. Engineers facing a vacuum of TBDs tended to fill it with expensive hardware and software. Marriott went on the attack. He declared the original, discredited concept study "the baseline." He called in engineers and made them explain why their particular subsystems had changed from the baseline. Usually there were good reasons—sometimes, in Marriott's view, there were not. Why were two $600,000 star trackers on the impactor? Unconvinced by the explanation, Marriott got rid of one.

He questioned the relationship between JPL and Ball. Marriott would have none of this "teamness." He read the whole thing as a smokescreen for "Brian Muirhead is in charge of everything."

Technically, Muirhead *was* in charge of everything. But Muirhead's tendency to delve into details, cultivated over years of in-house JPL jobs, translated poorly in dealing with a contractor like Ball. Marriott felt Muirhead was meddling in his business.

At one point, Muirhead came to Boulder, to the Tech Tower. Marriott, in his fifth-floor office, had a couch straight from the 1970s, bright orange with cherry wood trim.

"Brian," Marriott growled, "sit down on the couch. You and I have to have a talk.

"I respect the fact that you're the customer. You get to be the customer. I don't want to be the customer," Marriott continued. "But I run the program inside. I give direction to the folks inside. The folks inside are responsible to me. You need to deal with me. You and I will have a relationship. We'll get your job done. But I need to control costs. I need to control schedule."

Muirhead saw the partnership fraying. He battled constantly with the Ball business managers working for Jerry Chodil, the Ball Aerospace vice president, as they translated changes the spacecraft needed into contractual obligations that pushed up costs further yet. As money got tighter, the cultural rift widened.

He and Marriott "needed to be committed to each other's reasons for being in the partnership. My reason for being in the partnership is to deliver a scientific mission," Muirhead said. "John's primary reason to be in the partnership is to make a profit."

Carl Buck crawled between the trenches. To Ball, Buck was The Customer, at first not even allowed into Ball's Deep Impact computer network, but rather banished to one named "Outlander." Buck's JPL colleagues were also wary of him, suspicious he had gone native.

The mission burned through cash. Smoke alarms sounded at NASA headquarters. They sent a letter. *Come to Washington D.C. for a termination review,* it said.

In November 2002, Marriott, his boss's boss Jerry Chodil, Jet Propulsion Laboratory Director Charles Elachi, and project manager Muirhead found themselves in a NASA Headquarters conference room. Presiding was Weiler. Flanked by top deputies, Weiler sat at the head of a *U*-shaped table.

By that point, Weiler had cancelled five missions. Also weighing on him was the Mercury Messenger mission chosen the same day as Deep Impact. It was in even worse financial shape, en route to either a colossal overrun or his sword.

Muirhead and Elachi bore the brunt of the inquest. They made a presentation. Here's what's happening, here are the problems we see now, here's what it will cost to fix them. Some of the same NASA people who had chosen the mission grilled the JPL leaders.

Marriott kept his eye on Weiler. The man was a straight shooter, but volatile, hot-headed. What was he thinking? Weiler's deputies seemed ready to bury this thing. If Deep Impact had a prayer, it would hinge on a father's mercy.... Marriott made a mental note to never again be in a cancellation review. He liked being in control. He was not in control here. Then Weiler looked right at him.

"*You* tell *me* why I shouldn't cancel this program," he said.

The words hovered in the air. Years of work and his company's shot at deep space hung in the balance with them. Marriott wanted Ed Weiler and the rest of NASA Headquarters on his side, not against him. He told the honest truth.

Deep Impact had spent more money than expected, no question, Marriott began. The mission was harder than anyone had imagined. Deep Impact's management hadn't paid enough attention to cost and schedule. Well, he was paying attention now. He recommended plugging NASA headquarters directly into the Deep Impact loop through weekly conference calls and other means—to put some supervisory power back in the hands of the ultimate Customer. This, he knew, would rankle JPL, but so be it.

Ball and its subcontractors were slated to consume about two-thirds of the mission's money in the process of building and testing the two spacecraft and instruments. Weiler needed to believe that Ball could get the mission back on budget. Though listening to what Marriott said, Weiler was more interested in how Marriott said it. Weiler considered himself an expert in sincerity detection. With a good look at body language and eye contact, "I can usually tell if somebody's rope-a-doping me," he said. "And if they come in and rope-a-dope me, then I cancel 'em."

Marriott, he decided, was not rope-a-doping him. Weiler sent him and the rest out of the room and then back to Boulder and Pasadena, to keep working toward the cosmic explosion he was so looking forward to.

One man's job was done, though. Like an angry diety, NASA demanded a human sacrifice.

"Everybody knew up front we were going to have to be both very clever and very lucky, and we were unfortunately neither," Muirhead said. "I was the price that was paid for not killing the mission. That was the deal that was struck. And I understood that. But it didn't mean it hurt any less."

Part III

Building a
Comet-Hunting Machine

Chapter 14

The Biological Metaphor

By October 2002, comet Tempel 1 had stretched its gravitational line. The sun took notice and began reeling it back in, winding faster and faster as it did. The comet was coming.

Even those unaware that their quarry was grazing the orbit of Jupiter knew one date was firm: July 4, 2005. Miss it and you either wait 11 years for Tempel 1 (Earth would be on the wrong side of the sun to watch the fireworks on the comet's next orbit) or find another dirty snowball. Either would mean delays and more money and, more likely than not, another comet mission entombed with Champollion.

Deep Impact had won a reprieve from Ed Weiler, but the team—now about 200 people at Ball, another 50 or so at JPL, and a handful at the University of Maryland—continued what was shaping up to be a marathon sprint. Their spacecraft was to launch in January 2004, just two years away.

In those two years, the redesigned Deep Impact had to morph from various components and design documents into two fully tested, cooperative and communicative robots. In the design phase, a single engineer with some decent software could construct an entire digital Deep Impact in a single day, scrap it and do the same thing the next day. Now hundreds—even thousands, if one included workers and subcontractors—were charged with building the actual spacecraft. It would take creativity, collaboration, smarts and skill. But bringing a spacecraft to life boils down to effort—a massively parallel process of building, testing, finding problems, and coming up with solutions.

Deep Impact emerged not from a single narrative, but rather countless vignettes involving a diverse cast. The following pages amount to an impressionistic portrait of a spacecraft and the people who built it, with the stories of a small minority of those dedicated to Deep Impact's success serving as broad brushstrokes.

•

SPACECRAFT EXIST IN MINDS, in computers, and on paper for so long they can seem like literary creations. Most spacecraft remain imaginary. Deep Impact's transition to the physical world was heralded by the whirring and grinding in the Ball Aerospace machine shop.

About 20 men worked on standard mills and other machines turning blocks of titanium, aluminum, stainless steel and invar (a nickel-steel alloy insensitive to temperature swings) into hinges for solar panels, mounts for space telescopes, brackets, braces, boxes, and manifolds. They focused on the hardest, most complicated and precise parts of a spacecraft, farming out the rest to a dozen shops Ball continually vetted and certified. Machinists double- and triple-checked engineering drawings hung on the Plexiglas between their workstations, then programmed computers that commanded robots wielding sharp objects. The computer-controlled machines imposed their will, showering the concrete floor with silvery shavings as they turned metal blocks into spacecraft parts.

Gone were the cheesecake photos from the days of the University of Colorado Rocket Project, but the challenge lived on. Building for space meant perfection, or as close to perfection as a man with a machine tool could get. A screw-hole shifted by the breadth of a human hair could scrap a week's work.

Their parts were cleaned, anodized, inspected, checked and cross-checked, then sent over to the flight assembly shop, which had milling machines, too, but fewer. There, piece parts came together as components, which would be mounted to the spacecraft in mechanical assembly.

Hundreds of people worked on the spacecraft, but only a handful touched the flyby and the impactor. Joe Galamback and Alec Baldwin were among them.

Galamback, in his mid-thirties, was tall and fair, with the delicate hands of a pianist. In a suit, he looked like a middle manager, but he had once made his living in some of Colorado's grittiest machine shops. Galamback had been a job-shop mechanic repairing axles and grinders worn and broken in rendering plants east of Denver, where animal parts became pet food. He came home covered in grease and oil and solvents whose stench scoffed at detergent.

Better jobs followed, and when Galamback had the chance to move to Ball Aerospace, his wife Donna told him: "If you were pushing a broom there, you would be better off."

He started in the Ball machine shop in January 1996. He saw that no one wore jackets at their stations. He had never worked in a shop with climate control. He sniffed no hint of smoking oil, in his mind as essential in a machine shop as beer at a Denver Broncos game. Rather than oil, which reeked and smoked in the frictional inferno imparted by a drill bit on metal, this shop lubricated with alcohol—only alcohol, he was told.

Galamback was an experienced hand, but the shop foreman had started him slowly on brackets for the housing of a Hubble instrument. He was working away and thinking little of his labor's consequences when he wandered into a testing area where the finished Geosat Follow-On satellite—the erstwhile model for Deep Impact—waited in all its splendor to be shipped off.

"When I looked at that, that's when I freaked out," he said. "That's when the realization that what I'm making is going into space hit me."

He walked back into the shop and, his concentration shaken, made a mistake that turned the part he was working on into a hunk of scrap.

Galamback became the lead technician on the team physically constructing Deep Impact. It was a big job. He and another tech, Mike Haddox, sought help. It was like picking out a puppy, Galamback remembered. They found a friendly, high-strung Canadian.

Alec Baldwin grew up on a farm in Hamilton, Ontario. He was five foot four and weighed 125 pounds, a stature friendly to horses, which he rode well enough in his youth to land a sponsor as an equestrian rider. With time, he gravitated to machining, moving from shops in Canada to Pennsylvania and South Carolina. Baldwin was making parts for the electronics industry—very precise, high-end work.

His wife was from Colorado. They came back for her brother's wedding and happened by the Ball Aerospace campus. There's where you should apply, she had said.

"Yeah, right," Baldwin answered, looking at the Ball logo on the sign. "I'm not working for a jar company. There's no way."

Galamback and Baldwin worked separately at first, on small teams that spent months building full-scale models of the respective spacecraft. Baldwin focused on the impactor; Galamback sawed plywood and screwed door hinges to fashion a mockup of the flyby spacecraft's strange pentagonal prism. They attached faux computer boxes and heaters and copper propulsion lines where the engineers said the real things would be—even a stand-in fuel tank the size of a water cooler bottle. The fuel tank didn't fit, a realization that more than paid for the entire effort. A universal engineering truth held that catching problems in design was ten times cheaper than finding them in production.

When the inch-thick aluminum honeycomb panels for the spacecraft arrived, Baldwin and Galamback started working together. They spent weeks with the panels, aligning and boring hundreds of holes and securing threaded inserts on which the spacecraft's many components would hang.

Joe Galamback with the Deep Impact spacecraft body festooned with "mass model" weights. This quasi-static load test made sure the spacecraft body could handle the burdens of future components.

Galamback and Baldwin came in at 6 a.m., worked until nine, then with Haddox took turns supplying breakfast: microwaved eggs, microwaved burritos, microwaved Canadian bacon. They screwed into the panels' many holes about 50 weights—"mass models"—in place of hardware still months from delivery. Then they rolled the whole thing to a vibration table to "shake the shit out of it," as Galamback put it, making sure the aluminum body was solid and the inserts held. Those lacking Galamback's clarity called it a quasi-static load test, in which stresses were precisely measured using about 100 accelerometers the size of sugar cubes. That test and others proved the structure was solid. Now they needed some real hardware to hang on their empty honeycomb frame.

A SPACECRAFT IS A FLYING ROBOT. Deep Impact would have no arms or legs, no head, heart, or stomach, and no consciousness. Still, the spacecraft was physically (in metal and wires and silicon wafers) and conceptually (in software) subdivided into systems and subsystems that were to serve similar roles as organs or limbs. Galamback and Baldwin had been focusing on the structure, the bones. Hanging meat on them and infusing Deep Impact with perception and intelligence involved hundreds of people in multiple organizations with many management layers and organizational matrices and, most importantly, teams. Most of those on the 17-odd Deep Impact teams had less obvious relationships with the spacecraft than did Galamback and Baldwin.

There were many teams busy working on instruments, thermal control, power, telecommunications, spacecraft computers, software, propulsion, attitude control and autonomous navigation. Despite their contributions to Deep Impact's physical form and function, they would never lay a hand on her. No less integral were the test team, the quality-assurance team, the mission operations team, and the launch-vehicle operations team, among others.

Building the creature itself, although a monumental task, was only part of the challenge. Unlike Victor Frankenstein, who designed, built, and set his Creature loose to be violent and misunderstood and tragic, the Deep Impact teams plotted their creation's entire life like micromanaging Greek gods.

Mike A'Hearn and the respective JPL and Ball project managers sat atop Olympus. A small group of their deputies handled big chunks of the whole. The man ultimately linking the sweeping engineering and planning efforts to the nuts-and-bolts work of Baldwin and Galamback was John Houlton. Houlton, Deep Impact's lead production engineer, described the spacecraft in biological terms.

"It's just like an organism," Houlton said. Seven major systems, many subsystems beneath them, forced to cooperate under human rather than Darwinian pressures.

His choice of metaphor was understandable. Houlton had studied biology and psychology at the University of Colorado in the late 1970s and had gone to work for a company that made kidney dialysis machines. One thing led to another and he found himself working as a production engineer. In 1984 he came to Ball.

Kidney machines demanded precision manufacturing, but that world paled in comparison to the world he was entering at Ball. With terrestrial products, Houlton said, you built 100 prototypes for insiders to hammer on, then 1,000 more for beta testers at large. With a spacecraft, you built just one, and it had to be perfect. "You get a 99 on a test and you get an A," Houlton said. "You get a 99 in aerospace and you flunk. Because your mission fails."

Houlton's job was to make sure everything came together for the guys turning bolts—that all the wandering streams of hardware made it to the river flowing to Galamback and Baldwin and onto the finished spacecraft. In construction terms, he was the general contractor to Bill Frazier's architect. Houlton went to hundreds of design-review meetings. Deep Impact's "drawing tree," as it was called, had 1,019 branches, each with parts like so many leaves. Houlton inspected each one, rotating it in his mind, thinking about its form and makeup. At times he felt like a lawyer representing Galamback and Baldwin and their hands-on colleagues. *How's it going to work? Is there enough access to get in there? Can we take it apart if we have to?*

But Houlton was a manager. What drove him was time and money—schedule and budget. And schedule drove him more than budget, because if the project got behind, it hemorrhaged cash. So Houlton noted a titanium part in a design and asked, Does it really have to be titanium? Will aluminum work? Titanium took three times longer to machine. That extra time could put a subsystem on the critical path—the bottleneck—and risk holding up the entire program, which would devour money as it marched in place.

At Houlton's side were quality inspectors whose job it was to approve each of the roughly 100,000 parts that were to coalesce into Deep Impact, whether made in Ball shops or bought elsewhere. They tested batches for strength and inspected hardware with micrometers and precision measuring machines. When something passed muster, it earned a BR-49 file—a pedigree.

If something went wrong in space, the part might be 50 million miles away, but the document was in a file drawer. "It's because you're only doing one. It has to be perfect. It has to have a pedigree," Houlton said. "You'd think: a $600 hammer, how can that possibly be? Come and give it a try. Just come and be in the trenches for a while and you'll know why. It's three bucks for the hammer and $597 for the paperwork."

BENEATH DEEP IMPACT'S HONEYCOMB BONES were strung a thousand wires, tested over months one at a time and bundled together into ropes to be anchored at six-inch intervals or risk being torn loose during launch. The wire harness was a brainless nervous system and a heartless circulatory system.

Power would flow through some of those lines. Deep Impact, metabolizing sunlight, needed no digestive system. It would be a kind of primary producer, a flying photosynthesizer. David Wilson led a team of about a half dozen whose job was to ensure the spacecraft had the energy to flourish.

Wilson's grandfather had been the first superintendent of Canyonlands National Park; his father, raised in Moab, went on to pioneer X-ray semiconductor lithography at IBM. Wilson, raised in the IBM town of Aramonk, New York, inherited his father's technical mind and his grandfather's outdoor bent. His indoor hobby was building trucks.

"I've got an acetylene torch, I've got an arcwelder, I've got a MIG welder," Wilson said. "I backpack, I rock climb, backcountry ski. Technology is not what blows my skirt up at all. The natural world is much more influential on me. So living in Boulder and having a career that allows me to live here and enjoy what I like to do is significant."

He gained a measure of notoriety at Ball when a security officer got word of someone burying something furry in a snow bank. A uniformed man showed up at his desk, walked him outside, and asked him to explain the corpse.

Wilson had found a roadkill fox and had wanted to preserve it, he explained. He planned on skinning the animal and harvesting the fur to garnish trout flies.

He earned a master's degree in electrical engineering, came to Boulder with a girlfriend in 1987, and got a job with Ball. He migrated to power systems. Marriott brought him onto Deep Impact. Not long after, Wilson sat in one of the team-building workshops in the mountains above Boulder. The team-building consultant said, "If you were really committed to doing this mission, you would do it for free."

Wilson raised his hand.

"I'm actually a mercenary. You need to pay me. I will fight to the death. But I need to be paid," he explained. "I don't want to be on the team. I want to be a subcontractor. Tell me what you want done and I'll go out and do it."

He thought of asking the consultant: "If we didn't pay you five grand to show up and give these talks, would you come?" but refrained.

Power—solar panels, batteries and the connections between them—was an example where deep space made a difference. In Earth orbit, the daytime sun intensity changes only about 3 percent with the seasons. With the comet mission, at the moment of impact, the spacecraft would be 50 percent farther away from the sun than Earth. At that distance, Deep Impact's solar panels would produce just a quarter of the wattage of a similarly outfitted Earth orbiter. Without careful planning, the solar array could cook the spacecraft's electronics while close to Earth—or generate too little juice to power the flurry of activity during encounter with the comet.

The solar panels were black wings clipped at their outer corners so that they would fit in the rocket's nose with less than an inch to spare. They looked like overgrown versions of rooftop solar panels, but other than working by harnessing electrons smashed free of their semiconducting surfaces by the sun's photons, they had little in common.

Spectrolab—the same company that had supplied the first Orbiting Solar Observatory's solar cells—built the $2 million Deep Impact panels. Rather than a single layer of silicon, the solar array was a triple-decker sandwich of various gallium and germanium compounds, each picking off particular colors of sunlight and converting 27 percent of the energy into electricity—about three times that of a rooftop solar panel. The panels were radiation-hardened to the point that they would have withstood the OSO-crippling nuclear blast and, more relevant to Deep Impact, could shake off high-energy bombardment from the sun and cosmos.

The nearly 3,000 cells making up the panels were wired together in very particular ways to keep the power steady despite the ebb and flow of the sun during the mission. Such wiring, combined with commands from homemade power-control software and mission controllers' tilting the spacecraft slightly more while flying closer to the sun, would have the effect of shrinking and growing the solar panels depending on Deep Impact's location.

Batteries would store the power. With constant sun, the flyby could mostly rely on its solar panels. The nickel-hydrogen battery Wilson's team chose for the flyby would be no more powerful than the first OSO's battery.

The impactor battery, on the other hand, had to pack enough energy to run a 200-watt machine for 24 hours—demand that would drain a fresh car battery in less than four hours. Wilson considered a single, big lithium thionyl chloride battery for the job. Such storage devices, related to the lithium-ion batteries in cell phones and notebook computers, pack huge amounts of energy in small spaces. But this particular stripe of lithium battery had incendiary tendencies so pronounced that manufacturers kept them out of consumers' reach. A battery big enough to power the impactor's solo ride would, until fully assembled, be so volatile that it would have to be built by robots 500 feet underground in France. Wilson passed. Rather, he and his team strung together 216 smaller batteries, D-cells used in military radios, of the same potent chemistry. Those cells were made in France, too, but not by robots of the netherworld.

ONE OF THE POWER SYSTEM'S most demanding customers was the thermal-control system, where deep space also made a big difference. A spacecraft maintains its body temperature like a rabbit. Both creatures generate heat and then hold it in or dissipate it depending on the weather. Bunnies have fur and big ears. Spacecraft have heaters, blankets and paints.

Thermal engineers, like power engineers, had to contend with the 50-percent swing in solar energy no Ball spacecraft had ever contended with. Charged with figuring out temperature was John Valdez.

He was an applied mathematician, a helpful background for someone who spent his time concocting numerical "heat maps" of oddly shaped spacecraft whose many systems warmed and cooled at various times under diverse conditions.

Valdez had been the second person in his family to graduate from college. The first had been Harold Montoya, who happened to be a deputy to John Marriott on the Deep Impact team. Montoya was one of Valdez's 38 cousins. "Maybe it's not that unusual that two of us ended up at Ball," Montoya quipped.

Valdez's parents had driven to Denver from New Mexico with their young son and the rest of their worldly possessions in the car. They moved into low-income housing and eventually joined the U.S. Postal Service. Valdez's mother sorted mail; his dad carried it.

Ensuring that Deep Impact's perishable contents neither melted nor froze was the son's mandate. Similar to those worrying about temperature on the first OSO, Valdez spent time in front of a computer, a desktop PC thousands of times faster than the one Ball's thermal-control pioneers had had to feed punch cards.

Spacecraft thermal design was about maintaining balance in extreme conditions. Valdez had to know where the spacecraft was pointed and what it was doing through the entire mission. Every degree of rotation changed the sun angle and temperature mix; a flurry of activity by an instrument would pour electrons through various boxes inside the spacecraft, which would heat them up. Some systems liked to be warm—hydrazine-filled propulsion tubes were to be kept above 50 degrees (10 C), for example—and others, such as the big telescope's infrared detector, were more comfortable at minus-307 (-188 C). The impactor spacecraft complicated matters: like a baby kangaroo, it was to spend months inside the toasty pouch of the mother ship and then suddenly hop into the cold.

The spacecraft design was generally kind to Valdez. The solar panels would shade most of the spacecraft body like a parasol. The challenge was fighting the chill. Deep Impact's version of radiating rabbit ears would be an absence of insulation. The spacecraft's internal warmth could then radiate through into space. Over particularly hot items such as batteries, silver Teflon film would wick away yet more heat.

Keeping most of the spacecraft reasonably warm involved a variety of tricks. Valdez and his small team worked with propulsion engineers to design plastic insulators for propulsion tube mounts to block outside chill from reaching the tubes via the spacecraft's aluminum skeleton. They worked with mechanical engineers on designing special titanium brackets for the solar panels. Strength wasn't the issue; the brackets holding the sometimes boiling-hot panels would channel some of their heat into the spacecraft body, which Valdez wanted to avoid. Titanium conducts 10 times less warmth than steel and more than 20 times less than aluminum.

Unlike the OSO team, which had only paint, Valdez had electric heaters, thermal blankets and Kapton film in his arsenal. Deep Impact's network of 150 paper-thin, aluminum-etched plastic heating pads—roughly 100 in the flyby, the rest in the impactor—demanded up to 190 watts, seven times the juice the first OSO could have delivered. Valdez orchestrated the heaters' actions via an array of sensors feeding into a computer board dedicated to turning heaters on and off like the lights of an extreme holiday display.

Thermal blankets had been a product of the plastics revolution. Called multilayer insulation, or MLI, they were just an eighth of an inch thick. At first glance, they looked like chrome interspersed with lace. Closer inspection revealed 10 layers of aluminized Mylar—the stuff of Pop-Tart wrappers—each separated by mosquito nets of Dacron. The netting kept the Mylar leaves from touching and thus provided an insulating barrier. The blankets, cut and shaped by in-house thermal-blanket "seamstresses," were Velcroed and taped to Deep Impact, then overlaid with tough sheets of Kapton. Two hues of Kapton would cover much of the spacecraft: black to absorb heat over most of the structure, and gold over the instruments to reflect sunlight back into space.

THE PROPULSION SYSTEM, protruding into the cold and so prone to frostbite, was among Valdez's great concerns. For Steve Sodja, propulsion was paramount, and deep space made a difference here, too. Sodja was the only full-time engineer on the Deep Impact propulsion system, just as Dick Woolley had flown solo on the first OSO. Born the year OSO launched, Sodja had studied mechanical engineering and gone to work for Hercules Aerospace in Magna, Utah. He went to the Allegheny Ballistics Laboratory in West Virginia, which Hercules ran for the Navy. His work there had been scattershot—modeling diesel-engine combustion, designing solid-rocket motors, and shooting ordnance from tank guns.

Sodja went from smoke to mirrors in 1995, when Ball hired him as a design engineer specializing in mounting instrument optics. Following a stint as the lead mechanical engineer for the star camera used to aim the Chandra X-Ray Observatory, he joined Deep Impact. It was back to propulsion, but not the solid rockets he knew from his time at Hercules. Solid rockets were rubber-burning roman candles like the space shuttle's flanking boosters. On Deep Impact, fire-and-brimstone rocketry was Boeing's domain. Rather, Sodja was in charge of the spacecraft's small thrusters. Thrusters, the closest things

Deep Impact had to arms and legs, would let the spacecraft move around in the general confines of the trajectory Boeing's rocket imparted. Sodja's work, like Boeing's, was about thrust—generating pressure. "It's the same equation," Sodja said. "But everything else is different."

He helped whittle down the flyby's liquid hydrazine propulsion system from 20 thrusters to eight. He also looked hard at the nitrogen-gas system planned for the impactor. True, Ball had had success with nitrogen "cold gas" propulsion systems evolved from the one Woolley designed for the OSO 1. But the last of these had flown 25 years ago.

Hydrazine had become the propellant of choice despite its drawbacks. The chemical was extremely poisonous, for one thing. But then, hydrazine systems were reliable and produced more push per pound than cold gas. And while Ball had pleasant memories of nitrogen-gas propulsion, JPL did not. A speck of dust stuck in a valve can drain a propulsion system. It had killed JPL spacecraft. One such mission had been an early-1990s Strategic Defense Initiative satellite called MSTI. Its nitrogen leaked away in four days.

Ball had proposed nitrogen not out of nostalgia, but rather because the impactor lacked what had become a standard technology for orienting spacecraft. Called reaction wheels, their orchestrated spinning could rotate a spacecraft precisely and then lock it in a given orientation. The flyby spacecraft had four reaction wheels.

The wheel-less impactor had to maintain its orientation through a careful arrangement of gas impulses. The problem was, big impulses would yo-yo the spacecraft back and forth, smearing images the spacecraft's autopilot needed to aim for the comet. The impactor could only tolerate faint puffs.

Hydrazine's propulsive strength was thus also a weakness. Gentler nudges meant tinier droplets of hydrazine, which required faster valves. Valves for a hydrazine propulsion system for the impactor would have to open and close within about six milliseconds, or thousandths of a second. A human blink takes about 100 milliseconds; a traditional television screen refreshes every 30 milliseconds; a car on the highway covers six inches in six milliseconds. Sodja had never heard of thrusters so fast.

Sodja talked with experienced hands at JPL. Reaction wheels, although sealed, were moving parts, and moving parts broke and wore out. JPL avoided them wherever possible on its longest deep-space ventures. Neither the twin Voyager solar system probes nor the Cassini Saturn orbiter had reaction wheels. Both had needed fast thrusters. Sodja might check in with Aerojet, a propulsion person at JPL suggested.

Aerojet, spun off from JPL in 1942 to make rockets to help lift heavy bombers into the sky, was best known for liquid-fuel main engines of the sort powering the Boeing Delta II. But Aerojet also worked with hydrazine, and indeed still produced a version of the "minimum impulse thruster" they first developed in the 1970s for the Voyagers' grand solar system tours. With hydrazine thrusters capable of such butterfly burps, as Sodja called them, the nitrogen cold-gas system would remain a relic of Ball's past.

IF SODJA'S THRUSTERS DID THEIR JOBS, they would put two spacecraft in harm's way, one irrevocably. Both needed to survive the comet's abuse, including the impactor, which mission leaders hoped could withstand Tempel 1's barrage long enough to take a few good photos and lower its copper helmet into the comet's chest.

The risk of loss was serious. The probability of a comet grain pounding the flyby spacecraft was 50 times greater than that of an Earth orbiter suffering a meteoroid hit. At a relative speed of 10.2 kilometers per second, a BB-sized particle would deliver the punch of an exploding hand grenade. The nuclear attack on the first OSO notwithstanding, no Ball spacecraft had run into anything so severe as a comet's wrath. Even JPL's august resume was light on ballistics experience. Fortunately, one of the world's foremost experts in space bulletproofing worked right down the road from Ball.

Joel Williamsen ran the University of Denver Research Institute's Center for Space Survivability. He had begun his career designing warheads at the U.S. Army Missile Command. He moved on to NASA in the late 1980s, where he applied his expertise in hypervelocity destruction to preventing space junk from putting holes in the International Space Station and its inhabitants. He earned seven patents and a PhD during his decade at the Marshall Space Flight Center, his dissertation being a computer program for predicting the likelihood of meteoroid damage in orbit. "You create a geometric model of the spacecraft, you throw millions of particles at it, and you figure out what the overall likelihood of a penetration is," Williamsen explained.

In 1998, he salvaged from the NASA Marshall bone yard a high-powered air gun used to test Skylab shields and headed to Denver. He set up the gun in a Quonset hut on the plains 10 miles east of the Denver exurbs and hung out his shingle. The gun would lure clients, he figured; the computer modeling would keep them.

Deep Impact's redesign had done away with the 128-pound debris shield and left no obvious room for a replacement. Guaranteeing survival would have involved launching an armored truck. A'Hearn and the engineers settled on a 95 percent chance of the flyby spacecraft making it to the point of closest approach and a 90 percent shot at surviving the rest. The impactor had to make it through its last trajectory correction seven-and-a-half minutes before impact and be healthy enough to adjust for trajectory-altering comet particle hits after that.

Williamsen suggested spot shielding, which could save tens of precious kilograms. Spot shielding would be something akin to a soccer player's shin guards. But Deep Impact had many shins, each with its own unique vulnerability.

Fred Whipple, best known for his "dirty snowball" comet theory, also happened to have invented space shielding—in 1947, a decade before Sputnik launched. It consisted of two or more thin metal sheets. The first, a "bumper," would break an incoming projectile into countless fragments and molten globs; the second sheet would stop them. For larger projectiles, one would add more sheets of metal.

Lorna Hess-Frey sent over Deep Impact design files, which Williamsen and his colleagues used as input into a much-evolved version of his PhD work. The University of Denver team added digital shielding to critical areas—batteries, thrusters, electrical lines—and in some cases proposed changes to the Deep Impact design to incorporate shielding into the spacecraft. To keep aiming at the crater and the post-impact debris cloud while racing through space, the flyby spacecraft would expose six of its seven sides to comet particles. Among other changes, Williamsen suggested thickening up the honeycomb panels of two of the spacecraft's five side panels.

Aluminum and steel shielding being well known, Williamsen took aim at two unorthodox shielding materials. The first consisted of three copper sheets proposed for the leading edge of the impactor. They would be the first copper Whipple shields ever flown. The second was first-of-its-kind graphite epoxy shielding for the leading edge of the solar panels, chosen because aluminum would be too soft to help stabilize the panels.

In the plains east of Denver, Williamsen and colleagues loaded the impactor shielding—three layers of copper plus two layers of aluminum to simulate the impactor's exoskeleton—into the big gun's receiving end and evacuated the building for a neighboring hut. On the building's roof, a red light spun and a siren howled. A quarter mile up a dirt road, a sign read: WARNING: DO NOT PROCEED DOWN THIS ROAD WHEN RED LIGHT IS FLASHING. The two-stage light gas gun blasted a nylon slug into its target at 7 kilometers per second—16,500 mph. The top layers had no chance. But at bottom, the shields held.

Three layers of copper and two layers of aluminum shielding after ballistic testing. The copper simulated the impactor's proposed Whipple shields, the aluminum its honeycomb exoskeleton.

Williamsen proposed bumpers of all sorts—aluminum, stainless steel, titanium, the graphite epoxy—and ended up with 289 different varieties of shield on the two spacecraft, more than double the diversity protecting the entire International Space Station. Just when it seemed like they had the spacecraft suited up, the science team changed the game.

Carey Lisse, a Deep Impact scientist from Johns Hopkins University's Applied Physics Laboratory, reconsidered his model of the storm expected from Tempel 1, and upped his estimate of the violence by a factor of four. Williamsen's team ran some numbers and saw that the spacecraft now had a 29 percent chance of losing some important piece of equipment—a telescope mirror, a propulsion line, a fuel tank—as it passed the comet. The shielding was supposed to keep the probability of such a loss below about 10 percent.

The spacecraft's weight had already been nailed down to the individual bolt and washer. Every spared gram would go into extra fuel. Williamsen urged Ball to add another 13 pounds to shore up Deep Impact's defenses. The system engineers made an exception to an increasingly draconian stance against added weight.

Lisse's theoretical musings on Tempel 1's temperament had changed a spacecraft, albeit slightly. Humanity's second-ever look at the core of a comet, courtesy of a spacecraft called Deep Space 1, would bring surprises of much greater consequence.

Chapter 15

A Comet Throws A Curve

Science had been an afterthought with the Deep Space 1 spacecraft, launched in October 1998. Part of a NASA program called New Millennium, the craft's aim was to test drive a dozen new technologies so future mission designers might have the confidence to use them. Deep Space 1 was tricked out with miniaturized transponders and instruments, refractive concentrator solar arrays, experimental autopilot software, and even ion propulsion. Leaders of low-cost science missions such as Deep Impact tended to stick with proven technologies to keep price tags down. But such collective risk averseness slowed progress and threatened to make the U.S. space program more expensive and less capable in the long run. New Millennium Program technology was to play harmony to the Discovery Program's science.

Deep Space 1 had wrapped up its technology work by September 1999. But, its handlers argued, the spacecraft was already in deep space: why not do some science, as long as it was there? And so Deep Space 1 revved up its ion engine and headed for the asteroid Braille and the comet Borrelly.

By the time it approached Borrelley, the spacecraft was, in the mission leader's words, "an aged and wounded bird."[1] Deep Space 1 had been battered by solar storms; it had dribbles of fuel left; its experimental camera was standing in as a navigation device; it had no shielding against comet particles.

Deep Space 1's view of the comet Borrelly

There seemed scant chance that the spacecraft could contribute to comet science.

But on September 22, 2001, Deep Space 1 snapped its photos and survived to send its bounty home. The world had its second-ever look at the body of a comet, Halley having been the first. Even the best shot of Borrelly was of mostly black space, in which hovered, just north of center, a 50-pixel-wide, slightly blurred white-and-gray glob. Though it appeared to be about half Halley's size, five miles long and half that in width, Borrelly looked nothing like its more famous cousin. Of particular note was a distinct depression—almost a kink—in its middle. It wasn't quite a banana. Borrelly looked like...a bowling pin.

The world took scant notice, its attention swamped by the devastating terrorist attacks of September 11, 2001. The comet Borrelly also lacked the mystique of Halley in the popular imagination. Plus, the combination of Deep Space 1's distance from the comet and the limits of its experimental telescope combined for images less than breathtaking to those outside the tight community of cometary scientists. One notable exception was the Deep Impact navigation team.

A handful of people worked on Deep Impact's navigation, an all-JPL effort. The two most central to the effort were Dan Kubitschek, who led the autonomous navigation team, and Nick Mastrodemos.

Mastrodemos, in his late 30s, had grown up in the village of Thisvi in central Greece, the son of a midwife and a town administrator. Though familiar enough with the work of Sophocles and Aristophanes, as a boy he preferred *Star Trek* and *Lost in Space*. He ended up in a boarding school in Athens and then at Virginia Tech for a master's degree in physics. Realizing his passion was in astronomy, he earned a PhD at UCLA for work modeling stellar winds from red giants.

He came to JPL in early 2000 and used his astronomy and math skills to plot the courses of spacecraft. For Deep Impact, Mastrodemos did both optical navigation and AutoNav, which would combine to lead the mission to its comet.

Optical navigation would guide Deep Impact for all but the last two hours of its 268 million-mile trip. To estimate the spacecraft's location, optical navigation used a combination of radio techniques (which gain insights from the Doppler-shift stretching and squeezing of radio signals) and data from the spacecraft's attitude control system. Navigators also had to figure out where the comet was going to be at the moment of impact.

Everything in space was a moving target. Planets were relatively easy to nail down. Don Yeomans's Solar System Dynamics Group at JPL knew where Mars was at any given moment to within a kilometer. But comets were different. The jetting of their own thawing ices and the pull of planets had them knuckleballing about their trajectories. The approaching spacecraft itself would have to track Tempel 1 from the moment the comet was bright enough for the big telescope to register it.

Deep Space 1's experimental autopilot software, AutoNav, was the basis for the Deep Impact system of the same name. Mastrodemos looked at the bowling pin of Borrelly and was, on one hand, elated. AutoNav 1.0 had worked, and colleagues would have more faith in Deep Impact's version of it.

Then he began to worry. If Borrelly looked like a bowling pin, Tempel 1 could look like a bowling pin, which Mastrodemos described as a "noncooperative target." A noncooperative Tempel 1 would reflect light unpredictably, rotate haphazardly, and befuddle the Deep Impact autonomous navigation system Mastrodemos was working on.

For most of the ride to Tempel 1, shape would make no difference. During the first three months, the comet would remain invisible to even the high-resolution imager. Then the coma would form a dot bright enough to awaken a single pixel in the big telescope's detector. Navigators could then aim the spacecraft as if toward a star, which would grow in brightness until the mission's final hours. Then size and shape would matter tremendously.

Rather than just recognize a comet, Deep Impact's AutoNav software would have to target a particular bright spot on the comet, and with such confidence that AutoNav software running on two independent spacecraft thousands of miles apart and observing the comet from different angles would choose the same bright spot.

Then there was the matter of timing. Both flyby and impactor had to know exactly when the collision would happen. The cameras on both spacecraft were to shoot rapid-fire immediately prior to and, with the flyby, just after the big event. Don Yeomans and his JPL team could only estimate the comet's location at encounter to within about 800 miles. It meant impact could happen two minutes earlier or later than their best guess. Deep Impact's instruments were like photographers told to shoot a downhill ski race. Start snapping too early and the memory card is full before the racer shows up; shoot late and you get empty slope. AutoNav had to use its smarts to reset the mission clock.

And here was Borrelly, looking like something knocked out of a bowling alley beyond Jupiter. Shut off the lights and shine a flashlight on a bowling ball. You'll see one bright spot. Do the same with a bowling pin, and it depends on how the pin is pointed. There might be one bright spot—or two of them. At some angles, it's two bright spots separated by a shadowy trough. That would be a big problem for Deep Impact. AutoNav, Mastrodemos had assumed, would work by averaging a "center of brightness" on the nucleus and aiming for it. With a sphere, or even a reasonable potato, such guidance would take you to the single brightest spot—*voila!* But with a bowling pin at just the wrong angle, AutoNav could average the two bright spots and lead the impactor to drill a divot in the pin's dark nape as the flyby spacecraft dutifully imaged impenetrable blackness.

Give a team of JPL engineers an elongated bowling pin of a comet and, courtesy of the Deep Impact science team, an estimate that Tempel 1 was three times longer than it was wide (half the size of Manhattan, the scientists guessed), and one soon had models of a comet with a 30-degree bend in its center—a true interplanetary banana.

For AutoNav, the implications were extreme. Shutting oneself in a dark closet with a banana and a flashlight may raise eyebrows, but it will absolutely yield a bewildering variety of bright reflections and shady troughs. What's a spacecraft to aim at? The impactor could—could!—smash into a wonderfully lit spot that its flyby partner would find entirely obscured.

So in addition to such software as "blobber" (to find the tiny bright "blob" of the comet in the detector) and "centroid box" (same idea, bigger blob), Mastrodemos invented for AutoNav something called "scene analysis."

If a space banana or donut or dumbbell or cucumber or whatever yielded two or even more shiny spots, scene analysis software would tell the impactor to aim at the one nearest to where it figured the flyby spacecraft would pass. Such a thing is hard enough to visualize, much less render in space-qualified code. Mastrodemos ran thousands of simulations with comets of various shapes. When his imaginary impactors hit the wrong bright spot or missed altogether, he adjusted and updated the code and ran his simulations a thousand times again—until finally it looked like AutoNav would work. But nobody, least of all Nick Mastrodemos, could be sure.

NAVIGATION WAS ONLY PART of the story. The navigation system was a passenger giving directions. The attitude determination and control system was the driver. It told the spacecraft when to turn gently and point with an accuracy of less than a hundredth of a degree—and when to leap boldly through space.

The attitude control team was based at Ball, 800 miles from JPL's navigation group. Such separation was risky. The Mars Climate Orbiter had suffered its fatal English-metric conversion error in a similar arrangement between JPL and Lockheed Martin. Among the ways JPL and Ball tried to avoid an unfortunate repeat was to divide the team itself.

At Ball, there were Charlie Schira and Lew Kendall; at JPL, there was Mike Hughes.

Schira was the Ball lead, in his early 40s and already a veteran attitude control engineer, having spent years at Ball as well as satellite makers Spectrum Astro of Phoenix, Arizona, and Lockheed in Sunnyvale, California. Kendall,

Mike Hughes, left

a decade younger, served as Schira's right-hand man. Dark-haired and broad-shouldered, they might have been brothers. Kendall was a second-degree karate black belt whose father had owned a print shop and farmed 55 acres as a weekend hobby. He had graduated from Penn State and come to the University of Colorado for a master's degree focusing on control theory as applied to variable-speed wind turbines. Along the way, he taught classes. A senior Ball engineer brushing up on sophomore-level dynamics was impressed. Not long after, Kendall found himself at Ball.

Hughes, about Schira's age, was also a Penn State grad with ties to the printing business. He was a bit shorter than his Ball counterparts, with blonde-dyed hair and an unmistakable intensity. He had landed at JPL through an extraordinary series of events.

Hughes was the youngest of nine children. He grew up in a blue-collar Philadelphia neighborhood. In high school, he worked for a print shop, doing business cards, letterhead and whatever else came through the door. High school ended; Hughes stayed at the print shop. With ink-stained fingernails, he played bass in a rock band called Third Uncle. Then the space shuttle Challenger blew up on January 28, 1986.

Hughes was 23 at the time. He had always been fascinated with space—Apollo, Carl Sagan, *Star Trek*—and the disaster shook him profoundly. Not that things were going well otherwise. His girlfriend had left him. He had lost his job. His car had broken down. His band had broken up.

Hughes decided he would go into aerospace engineering so he could "help NASA."

A *Time Magazine* story on Ron McNair, a man of humble origins who had risen to become an astronaut and who had been killed on Challenger, fueled Hughes's fire. A nephew spending the summer as a trash collector rescued curbside books on math, chemistry and physics and passed them on to

Hughes, who crammed and took Penn State's entrance exam and passed. The admissions people checked their statistics and gently mentioned he'd have a one-percent chance of graduating in aerospace engineering.

Hughes finished second in his class. He worked for GE Astronautics for nine years, most spent working on a Goddard Space Flight Center satellite called Terra. When a Goddard man asked Hughes what he was doing next, Hughes said he was going to JPL.

"You don't seem quite arrogant enough for JPL," the man said. Before Hughes could open his mouth, another added, "Oh, that's OK. They have a training program for that."

In Pasadena, Hughes went to work on an orbiter mission destined for Jupiter's icy moon Europa. As that mission limped toward cancellation, he interviewed with Brian Muirhead for a job on Deep Impact. He became JPL's lead and sole representative on the attitude control team.

The fundamentals behind attitude control hadn't changed since Dave Stacey mused about his "Ideal Control" for the University of Colorado Rocket Project. There were still "desired quantities," still "output quantities," still the need to unify them via "motive power." But the servo systems on the biaxial pointing control and the first Orbiting Solar Observatory paled in comparison to what Deep Impact's attitude control engineers were creating.

Unlike OSO, there were three spacecraft to control. The mated pair counted as one. Heavier and bottom-loaded, it would behave differently than the separated flyby and impactor spacecraft.

In addition, rather than relying solely on the brightest thing in the sky, Deep Impact would for months aim itself toward an invisible target. It was a complicated affair. For starters, ground navigators would use radio Doppler data and snapshots from Deep Impact telescopes to feed the attitude control system their best guess of the locations of comet and spacecraft, using the language of Chebyshev polynomials—or "Chebys"—after their Russian inventor. Then the onboard attitude control system would rely on its own eyes and ears. Star trackers served as eyes, providing spatial context and pointing accuracy; gyroscopes were its inner ears, lending stability.

Ball built the star trackers, two for the flyby, one on the impactor. Their job was to take snapshots of space and compare what they saw against an internal star catalog. Software gathered the star-tracker output and sent it to the gyroscopes. The gyros could sense if Deep Impact's position changed between snapshots. The brains of the attitude control system weighed what its eyes saw and what its inner ears felt and told the navigation system how the spacecraft was oriented in space. The cycle started again a tenth of a second later.

Sodja's thrusters were all the impactor spacecraft had for motive power. The flyby had thrusters, too, mostly for changing its orbit during the six-month cruise to Tempel 1 and then for fleeing the comet's path once the impactor was loose. But most of the flyby's motion—hundreds of planned twists and turns to aim its telescopes—would come courtesy of four reaction wheels. These eight-inch discs would "spin up," forcing the flyby spacecraft to move in an equal and opposite reaction. The spacecraft being so much more massive, it would turn gently, nine times slower than a clock's second hand.

That, at least, was how it would all work in theory—a perfect marriage between JPL navigation and Ball-based attitude control built into a clutch of high technology operating in synchrony. Reality would be less kind.

AT BALL AEROSPACE, there were instrument people and spacecraft people. They shared a friendly rivalry, a competition for primacy in their little corner of east Boulder. During the Orbiting Solar Observatory era, the spacecraft people reigned. With the instrument successes on Skylab, the Hubble Space Telescope, and dozens of other orbiters and even deep-space machines, the balance of power had shifted. Deep Impact, with instruments and spacecraft both, was emerging as a sort of tie.

Deep Impact's instrument crew was a diverse bunch. The manager, Marty Huisjen, had an experimental physics PhD in microwave technologies. With optics, he said, "the wavelengths are a lot shorter, you know, but the principles are not that much different." Don Hampton, the lead system engineer for the instruments, had shared an office with Alan Delamere's son Peter while the two worked on their PhDs at the University of Alaska. In Fairbanks, Hampton had built sounding rocket instruments to observe the northern lights. Delamere visited his son, the three cross country skied together in Denali National Park, and Delamere recruited Hampton to Ball.

The lead optical designer, Jim Baer, was the son of Ralph H. Baer, the man who invented not just a video game but *video gaming*. Baer joked that he was once the world junior Pong champion, having defeated his little brother on the planet's only video-game console. Tom Yarnell, the lead mechanical engineer responsible for mounting optics in precisely the way Baer envisioned, was the son of a Los Alamos nuclear physicist. Out of college he went to work for Hughes designing rotary drives. The company had specialized in "spinners," or spin-stabilized spacecraft like the Orbiting Solar Observatory. For lubrication, the Hughes satellites had relied on what Hughes engineers referred to as

"HAC Kote" (*HAC* standing for "Hughes Aircraft Company"). A Ball lubrication engineer had defected to Hughes with Marion Fulk's secret Vac Kote formula. It was all the same to Fulk, who had become a genius-in-residence at Lawrence Livermore National Laboratory.

To provide time for testing, Deep Impact's instruments were to be delivered long before the spacecraft shipped to Florida, so the team was under pressure from the project's earliest days. They maintained a sense of humor. In the Tech Tower, they worked on the fifth floor, directly above the spacecraft team. They printed a poster on an engineering-design plotter and lowered it outside fourth-floor windows. Lorna Hess-Frey, managing the spacecraft design team, was at her desk one afternoon as the sign inched down. It read, "INSTRUMENTS RULE, SPACECRAFT DROOLS."

The spacecraft team responded with a poster hung in the instrument team's work area. It had an image of the Deep Impact instruments mounted on a Greyhound bus, and asked, "ALL BUSES ARE THE SAME—WHICH ONE ARE *YOU* TAKING TO TEMPEL I?"

Such diversion aside, designing and building a space instrument was serious business, to say nothing of three instruments on two spacecraft. The bus would carry them to the comet, but the instrument team had to capture the moment: their mirrors and lenses bending and shaping light, their semiconducting detectors collecting electrons smashed loose in hailstorms of photons and arranging the resulting mess into ones and zeroes. Their custom software and computer boards would take this binary runoff and reassemble electronic currents into vivid images and spectra. Their work spanned optical physics, computer science, and electrical and mechanical engineering. Building space instruments was among the least-forgiving pursuits in engineering. Even the slightest slip-up invited disaster, as the Hubble Space Telescope had shown.

Baer and colleagues generated designs showing beams of light darting about optics in ways that gave the impression of unplayable string instruments. The flyby spacecraft had its High Resolution Instrument and a sidekick Medium Resolution Instrument, both classic Schmidt-Cassegrain reflecting telescopes with 11.8-inch (30 cm) and 4.7-inch (12 cm) primary mirrors, respectively. On the impactor was another medium-resolution instrument called the Impactor Targeting System. It would be, as Yarnell described it, "a ten-million-dollar disposable camera."

The big telescope was to be the most powerful ever to leave Earth's orbit. It could read a newspaper on the sidewalk from the top of the Empire State Building, or discern the difference between an SUV and a truck from across

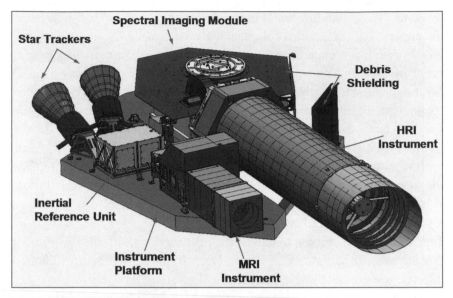

An engineer's rendering of the Deep Impact instrument platform, above, and the High Resolution Instrument telescope, left. Deep Impact's main telescope was to be the largest ever to leave Earth orbit.

the state of Colorado, or, most importantly, make out details less than five feet across in a freshly blasted comet crater 435 miles away. A prism called a beam-splittter parsed light coming in from the telescope, sorting through the stack of rays and tossing them at either the CCD visible-light detector or the infrared spectrometer, depending on the wavelength.

The infrared spectrometer, designed to determine the comet's chemical makeup based on the light coming from its surface and, if all went well, in-nards, had evolved from an instrument proposed for Delamere's 1996 Discovery mission to Venus that NASA never bought. The spectrometer bounced light off five mirrors and through two light-shaping prisms before funneling it into a sensor designed to capture redder-than-red wavelengths starting at 1.05 microns, or millionths of an inch. Human eyes can see from the violets of

0.4 microns to the blood-red 0.7 micron wavelengths. The detector reached all the way up to 4.8 microns, which people don't see but rather feel. Baer added touches such as a single prism that could both redirect and reshape light, saving space in the instrument as well as preserving a tiny bit more of the incoming beam, which each mirror bounce or plunge through glass diluted.

Operating at room temperature, the spectrometer would saturate like over-exposed film in a matter of seconds.[2] One of the instrument's key innovations kept the detector cold enough to operate. Refrigeration with liquid helium was out of the question—too complicated, too many millions of dollars. They would have to harness the cold of space. John Valdez, before his turn as Deep Impact's thermal engineer, had devised a double radiator to this end. One radiator wicked away heat seeping toward the chilly detector from the rest of the spacecraft; the second radiator, looking like a white dinner plate protruding three inches into cold space, conducted heat from the infrared sensor itself into the vacuum.

Temperature also played into a second surprise from Deep Space 1's swing past the comet Borrelly. In addition to snapping photos, the spacecraft had taken Borrelly's temperature. It found much of the surface to be a toasty 135 degrees.

To Mike A'Hearn, it was no surprise at all. That the surfaces of dirty snowballs could get hot had been common knowledge among specialists since Halley's 1986 visit, when Vega, a Soviet spacecraft, found parts of the famous comet's surface to be cooking at 300 degrees. Yet that knowledge had not trickled down to the team that was building Deep Impact's instruments. Should Tempel 1's sun-facing skin be as warm as Borrelly, a standard three-second infrared exposure would blind the big telescope's infrared detector.[3]

That, in turn, would negate the major premise of the mission—*to see what's inside* a comet. Only the spectrometer could peer into chemical makeup. Without its input, Deep Impact would bring home insight into a comet's texture, but nothing about its flavor. The mission could only hope the Hubble Space Telescope and other distant observatories could capture a few spectra.

In six harried weeks, Baer and his colleagues came up with, installed, and tested a fix—a sliver of millimeter-thin glass to be mounted across the middle of the spectrometer's entrance slit. It would turn away most of the hottest light like a good pair of sunglasses. Deep Space 1 could add bailing out Deep Impact to its long list of accomplishments.

•

SOME 30 PEOPLE WORKED on the instrument team, engineering the telescopes and a slew of supporting hardware and electronics. They enlisted subcontractors specializing in prisms, mirrors and sensors, carbon-composite baffles and various other components.

The instruments had their own computers on both the flyby and the impactor, with six electronics boards and associated software dealing with general operation, power supply, communications with the rest of the spacecraft, the light-detecting sensors, and the operation of filter wheels. Both flyby telescopes had such wheels, which were based on the filters A'Hearn had designed years before for ground observations of Halley. Each rotated one of nine filters in front of the visible-light sensor. Some filters let only certain wavelengths pass to highlight comet dust or key chemicals.[4]

The team worked at Tech Tower desks and in instrument labs, then later in a down-flow tent—a portable super-clean room—erected inside the clean room in which Joe Galamback and Alec Baldwin and others were slowly assembling and testing the rest of the spacecraft. Even the clean room wasn't clean enough for the instruments.

The clean room was in Ball Aerospace's Fisher Complex, where planning and design met metal, glass and silicon. The building had been named to honor John Fisher, the Ball Corporation CEO, who had retained his brother-in-law's enthusiasm for the container company's space business long after Ed Ball's retirement to a life of New Mexico ranching, Muncie philanthropy, and world traveling.

The Fisher Complex had to be one of the strangest buildings in the world. It wasn't the architecture, although various additions lent it the appearance of a building-block castle assembled by preschoolers. It was what was inside. On the third floor was a cubicle farm/office area the size of a supermarket. Below was a massive clean room, the length of a football field and half as wide, designed to host two space-shuttle payloads at the same time, a dream that had never materialized. Down the hall from the big clean room were the machine shop, grinding out parts; a quality assurance laboratory with microscopes and other inspection paraphernalia; and the equivalent of a massive torture chamber for spacecraft. This final space harbored vibration platforms capable of imparting bone-shattering force, various oversized Thermoses to freeze and thaw hardware, and a towering echoless room for bombarding spacecraft with radio waves.

Most of the instrument team settled into the Fisher Complex cubicle farm. Others, such as Jim Badger, a technician whose job on the instruments was akin to that of Galamback and Baldwin on the spacecraft, spent their days in the clean room. Yarnell, the mechanical engineer for the instruments, also spent hours with Badger. Yarnell liked to have his hands on the hardware, to the point that Badger referred to him as "my assistant."

Engineers were paid to design. On Deep Impact, design happened in cubicles or small offices or at home on laptop computers after the kids had gone to bed. Design did not happen in clean rooms. Narrowly viewed, time on the shop floor was time away from a design engineer's job. Managers scheduled such an engineer's time based on a certain volume of designs. The more time an engineer spent on the shop floor, the later at night (or earlier in the morning) that engineer would be at his desk doing his actual job.

But Yarnell loved the shop floor, watching mechanics and technicians build what he had imagined. How else could he have learned where his weaknesses lurked? In a sense, he was a throwback to Ball's earliest days, when David Stacey insisted his design engineers spend time in assembly to understand the hard consequences of their pencil-and-paper musings.

Yarnell worked prodigiously—sixty-, seventy-, eighty-, ninety-hour weeks. Though physically doughy, he was a cognitive ultramarathoner, capable of sharp focus hours after others slipped beyond caffeine's reach. His wife he described as "a saint" for putting up with him. His two daughters went for days without seeing him.

He dabbled in sculpture, did some painting, even tried out with a couple of colleagues for a TV show called "Junkyard Wars," in which contestants had a few hours to turn scrap heaps into boats or dragsters or rockets. Their tryout video, which involved acetylene torches, became a hit at team meetings. But engineering space optics was Yarnell's thing.

With Deep Impact's instruments, temperature was among his chief concerns. With falling temperatures, materials expanded or contracted in different ways. In the deep-space freezer, the carbon-composite telescope barrels would expand, for example, even as the titanium flexures connecting the mirrors to the barrels contracted. The aluminum connecting the telescope to the detector assembly would contract even more. The Zerodur telescope-mirror glass, though designed to be indifferent to temperature swings, would contract a bit and subtly warp. The cold grip of space could blind an instrument.

There was also the issue of gravity. The instruments had to be tested at 1 g on Earth, but would do their real work weightlessly. Moisture presented a problem, too. Even in relatively dry, climate-controlled Colorado air, remnant humidity soaked into the carbon-composite telescope barrel and made it expand. To complicate matters, the instrument had no way to adjust its focus once aloft—a focal mechanism had been deemed unwieldy and too expensive.

So Yarnell had to ensure that the instrument, badly out of focus at room temperature, contracted into perfection.

He made a spreadsheet—not a big spreadsheet, just a few hundred rows. He typed in daunting equations. When he was done, his spreadsheet told him where the mirrors should be mounted. Later, a subcontractor specializing in such thermal modeling used fancier software to confirm Yarnell's conclusions.

On the three telescopes, Badger and Yarnell mounted mirrors—concave primary mirrors to collect light from the comet and convex secondary mirrors to bounce it back through the donut holes of the primaries into the instruments. It was a painstaking process culminating with testing on a vibration table, where the sound of a telescope mirror cracking seemed to resonate across the Deep Impact mission, bringing with it all sorts of woe.

Chapter 16

Cracks and Contingency

Problems board spacecraft in nefarious ways. Some sneak on like ninjas. Others disguise themselves as trivialities, then band together with other trivialities and attack. Sometimes, a problem masquerades as a solution. This was the case with the big telescope's ruined mirror.

By the summer of 2002, mirrors, lenses, prisms, and filters had been mounted and aligned. It was time to test the integrity of Deep Impact's telescopes. The Ball instrument technicians placed their precision optics on a vibration table in the integration and testing area. Test engineers had once programmed "God's own subwoofer," as some called it, to play bad renditions of holiday carols.

Thirty seconds into the test came a "pow," then a gruesome rattling. There was no kill switch, so it went on for another 30 seconds as Tom Yarnell, Jim Badger and others exchanged glances with raised eyebrows. When they examined the instrument, they saw their perfectly polished, 12-inch mirror held in place by just two mounts. The third mount gripped only a shard of the mirror. The mirror had only three points at which the telescope could grab it, and one mount was now ruined. Yarnell was suddenly leading a review board of Ball engineers whose job was to come up with a strategy to present to management and The Customer.

"It's like, O.K., we have a problem," Yarnell later explained. "That's what we're good at. We'll find the problem, we'll figure out what happened, we'll figure out the solution, we'll go fix it. That's what Ball does. The basic design stuff isn't that hard. When you earn your pay is when you find something that's wrong and you figure out, O.K., how do we take care of this?"

Yarnell started working 80-hour weeks. One idea was to salvage the cracked mirror and craft a new mount to grab around the divot. The other was to use the backup mirror. Even tight-fisted Discovery missions double-ordered certain indispensable parts. But they couldn't just glue in the spare, put it back on the vibration table, and hope for the best. Losing the backup mirror to another crack would be catastrophic.

Yet how could they be sure the backup would survive? LightWorks Optics, which had made the mirrors, had suggested leaving the original mirror's mounting surfaces finely ground rather than polished smooth, and Ball's instrument team, after conferring with in-house materials experts, had agreed. But even the finest-ground surfaces harbor tiny cracks. Those fissures had proven fatal on the vibration table.

The new mirror mounts had to be smooth. Progressive polishing—using finer and finer grits until such microfractures were worn away—would have been the approach of choice. But the backup mirror's reflective coating was immaculate. Polishing the mounting spaces risked marring the coating. Another option would involve etching using an acid that ate away rough glass. Yarnell shuddered at the idea of dipping the mirror in acid.

He stumbled across something called Etchall. It was a product more familiar to readers of *Crafts Beautiful* than *Optics & Photonics News*. The acid paste was a favorite for etching words and designs into flower vases, fruit bowls, barbershop windows, and youth baseball awards. Yarnell was intrigued. He ordered a bottle.

He and Badger tested it. After 15 or so applications, Etchall was as good as an acid bath and less risky. Yarnell and colleagues packed up the primary mirror in a FedEx box and followed it to California, where they watched LightWorks Optics technicians apply Etchall to a $45,000 piece of glass upon which a $300-million mission hinged. Back in Boulder, they mounted the backup mirror in the big telescope and watched with trepidation as it revisited the vibration table. This time, it survived.

Yarnell's fix, simple as it was, had taken hundreds of expensive hours and risked a delay of weeks. While Deep Impact's managers hadn't budgeted for it, $50 million in "contingency" funds had been built into the project's bank account for just such surprises. The rainy-day fund could cover only a few

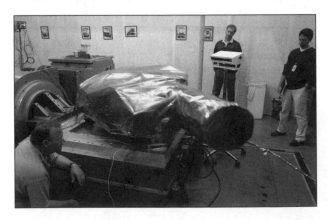

Tom Yarnell, left, Jim Badger and Monte Henderson look on as the instrument platform undergoes vibration testing following the replacement of the cracked telescope mirror.

more big things going wrong, and big things seemed to be going wrong at an alarming pace.

Not long after the mirror was fixed, the propulsion system was struck with a case of thinning, clogged arteries. Steve Sodja's system, with its snaking, stainless steel tubes, was to be among the first things mounted to the dual spacecraft. The impactor's tubes were already in place. The flyby spacecraft was within two welds of being done. Then a weld blew out.

These things happen. But when the Ball welder removed the section of tube, he noticed gray gunk inside. In hydrazine tubes, gunk could bring disaster. Sodja sent the stuff to Aerojet to see if it reacted with hydrazine. It didn't. Still, the goop could gum up a thruster, or microscopic particles could settle in valves and stop them from closing completely, causing fuel to leak away just as JPL had feared with a nitrogen-gas system.

At about the same time, during routine X-rays of welds, they found some of the propulsion-tube walls to be one-fifth as thick as they should have been. Deep Impact's vital maneuvering was suddenly at the mercy of polluted, thin-skinned propulsion tubes. The problem rose through the Deep Impact ranks, to JPL and to NASA Headquarters. Replace the tubes, the agency advised. So Sodja and Ball welders did, and it cost more time, more money, more contingency funds.

SUBCONTRACTORS TO BALL had their share of problems, too. One company, Starsys Research, was a Ball neighbor in Boulder. Starsys began in the late 1980s when an engineer named Scott Tibbitts started looking for interesting places to sell cigar-sized pistons filled with paraffin wax. He was in the water-heater business at the time. The wax, when heated, expanded and pushed the piston slowly, but with considerable force. Cooled, the wax contracted again. There were no explosive discharges, no gears to fail, no electrical complexities. Just wax and heaters.

Tibbitts talked with people in the solar industry, the medical-device business, even nuclear-energy people. He thought about space, where mechanisms—moving parts—were still notoriously unreliable. He flew out to JPL. Engineers there watched the wax-driven piston in action and told Tibbitts he might have something interesting there.

By the time Deep Impact rolled around, Starsys had landed mechanisms on more than 200 spacecraft. The company was supplying Deep Impact with hinges, bolt catchers, and struts to pop the solar arrays out after launch and hold them flat to the sun. The struts presented a challenge. The spacecraft needed lightweight springs to push the solar panels open after launch. Brian Buchholtz, a Ball mechanical engineer, suggested Stanley measuring tape. The yellow Mylar, when bent, wants to stretch back out and snap firm. Two strips of such tape, facing each other lengthwise like parentheses ( ) would do the trick, Buchholtz decided. Kinked nearly flat at launch, the tape would drive for linearity when the solar panels' bolts released, snapping the panels into place.

A Ball Aerospace manager popped into Starsys one day on other business and noticed something on a conference-room table.

"What's this?" the Ball man asked.

"It's a gimbal," Tibbitts said.

A gimbal is a sort of mechanical wrist used to point an antenna. This particular gimbal was about the size of a submarine sandwich.

Deep Impact engineers had debated whether to use a gimbaled antenna or just rotate the entire spacecraft every time it was to communicate back to Earth. The Live for the Moment strategy made the gimbal indispensable: Deep Impact had to shoot imagery and transmit at the same time. But money was tight—Ball could spend only about $800,000 on a gimbal.

It was about the biggest contract Starsys, with its 70 employees and about $6 million in annual sales, could handle. The company had done similar hardware—the gimbal on the conference room table, for one. So accepting a fixed-price contract, where Ball would pay only so much no matter what it actually cost to make, seemed a reasonable thing to do.

As the specifics of what the gimbal was expected to do trickled into Starsys, they realized they had erred. Given the powerful high-gain antenna's narrow beam, the Deep Impact gimbal had to point back home with an accuracy of a quarter degree. But the gimbal would also have to brace the meter-wide

The Starsys gimbal

antenna during launch. Combining such strength and subtlety into a single mechanism was difficult. The gimbal also had to be extremely "quiet," moving gently so as not to jostle the spacecraft and blur comet images.

"All those things got reduced to requirements, but we are just thinking we are going to make this thing the size of a Subway sandwich and it is going to cost a million dollars," Tibbitts said. "So when you review the requirements document, you have something simple and small in the mind, and it filters your review. And in the end you realize somebody should have stood up and said, 'This is *not at all* what we envisioned at the beginning.'"

The Deep Impact gimbal would demand exotic, expensive materials and electronics well beyond anything Starsys had ever done. The company was building something totally new on a tight, fixed-price deal. The Subway sandwich grew into something the size of a swaddled baby. The payments from Ball remained mercilessly flat. In retrospect, Tibbitts said he should have stood up and said, "Stop. This is a very different duck." Starsys spent more than a million dollars of its own money to get it done. The contract put Tibbitts's entire company at risk. There were delays. Ball engineers spent more time at Starsys, and engineers cost money. Away went more of Deep Impact's "contingency."

GIVEN THE POINTING ACCURACY required to take aim for a comet, Deep Impact needed the best gyroscopes available. They were so sensitive that traffic on a nearby road could throw them off during testing. The gyroscopes did not spin, but rather vibrated, containing "wine-glass resonators," so named because of their hemispherical shape. Litton Industries made them in Goleta, California. Then Northrop Grumman made them 80 miles away in Woodland Hills. Therein lay the problem.

Grumman bought Litton in April 2001 and moved Litton's Guidance & Control Systems Division to Woodland Hills. The idea had been to finish building the Deep Impact gyroscopes in Goleta and test them in Woodland Hills. Had the gyroscopes sailed through testing, the story would have ended there.

But before the transition, the angle programmed into a single welding machine had been off by 15 degrees, and the resulting welds—to mount the fragile wine-glass resonators on the assembly holding them—penetrated about a tenth of a millimeter deep, less than half of what they should have been. A sheet of paper too shallow, roughly. The error affected several test gyroscopes, as well as ones slated to fly on Deep Impact and Mercury Messenger.

Grumman subjected the test gyroscopes to a nasty vibration test (nearly two-and-a-half times as violent as Deep Impact flight units would face), and they snapped. The welds broke again on a test about 50 percent rougher than Deep Impact would demand. The Deep Impact units, facing lesser stresses, passed. They were probably fine, Grumman said.

Probably. Ball had its reputation as a player in deep space riding on this mission. JPL was equally cool to the idea of losing a spacecraft to a botched weld.

A cracked gyroscope weld at 200 times actual size

Complicating matters, the only way Grumman had known the test gyroscopes' welds had snapped was by dissecting them and having a good look with a scanning electron microscope. The Deep Impact gyros were sealed and ready to go.

A "tiger team" of 10 engineers from JPL and Ball, including attitude control people, materials experts, and a metallurgist, went to work. Grumman said the welds met requirements; Ball and JPL insisted otherwise. In other circumstances, they would have pressed Grumman to rebuild the gyroscopes. But the move from Goleta to Woodland Hills changed the calculus.

Eighty miles was a long commute anywhere. With greater Los Angeles traffic, it was unbearable. Some of Litton's experienced hands decided not to make the move. Those who did would be working on high-precision machinery just unbolted and schlepped down the highway and set up again, which could introduce new quality problems. Deep Impact's gyroscopes may have been flawed. Would new ones be any better?

Engineers from Pasadena and Boulder paid repeat visits to Grumman. NASA Headquarters again got involved. The tiny welds became a huge deal. Perhaps three months into the analysis, John Marriott, the Ball program manager, decided he had seen enough. They could argue that the Deep Impact gyro welds were no good. But they could argue that the welds on rebuilt gyroscopes would be no good, too. This former quality-assurance technician knew a bad weld when he saw one. But he figured that, with help, the welds would hold up during launch.

For that help, Marriott again called on Yarnell. While Yarnell and a small team worked for weeks on ways to buffer the boxes holding the gyros from launch vibrations while ensuring their rigidity once in space, which was vital to navigation. He experimented with packing polyurethane goo into the joints of thick titanium S-shaped flexures holding the gyro boxes. It wouldn't be enough, tests showed.

The team came up with an elaborate platform made of aluminum and polyurethane. On a vibe table, with a simulated gyroscope, the fix seemed to work and the problem was deemed to be solved. But fires burned elsewhere on Deep Impact. Misbehaving electronic pulses in flight computers became the mission's gravest threat.

•

THE DECISION TO UPGRADE Deep Impact to faster RAD750 processors had meant brand-new computers—two for the flyby and one for the impactor. The Southwest Research Institute in San Antonio, Texas, had won the contract to build them. SwRI ("swih-ree," as the nonprofit was known) employed 2,500 people doing contract research and high-end engineering on everything from car engines to biomedical devices. SwRI was what Ed Ball and Art Gaiser had dreamed Ball Brothers Research might become. On Deep Impact, SwRI's 60-person Space Systems department became a subcontractor to what Ball's Boulder offspring actually grew up to be. About 45 people designed and built the Deep Impact computers in San Antonio. A core group of 10 engineers and technicians worked on them six days a week for the better part of two years.

There were hardware problems. Field-programmable gate arrays—chameleon computer chips capable of reshaping themselves for particular processing tasks—misbehaved. Semiconductors malfunctioned, connectors snapped, and the memory boards got over-amped and had to be replaced. Making computers fast enough to deal with the huge data throughput the Live for the Moment strategy demanded would prove to be harder than anyone imagined. To boost speed, engineers designed computer hardware to do the job that more flexible but slower software might have handled on a less-demanding spacecraft. But speed was only one aspect of the challenge.

"The biggest difference between our computers and the one on your desk is we don't have the grand reset switch," explained Michael McLelland, a SwRI program manager. "So when you get the blue screen of death, you can't just hit the button and reset it because you could lose control of the spacecraft."

So the spacecraft computers were redundant physically—the flyby had two complete boxes—as well as logically, such that the boards themselves could process along parallel paths should a cosmic ray temporarily blind a circuit. Cross-strapping allowed one computer to take over communications or image processing. Flexibility added complexity.

"There's a rule in this business that the more fault tolerance you have, the harder it is to test and finish," McLelland said. "What happens is every test I run, I have to run on both the primary and the secondary. And so if I'm two days into a test and I find a problem, I have to start all over again, and I have to have both boxes working exactly the same. That's where it really gets to be time-consuming and tedious and just difficult to deal with."

That hurt SwRI, which, like Starsys, had agreed to build the computers for a fixed price. On one hand, the fixed price made sense. Deep Impact was one big fixed-price contract, given the Discovery Program cost cap. On the other, etching a dollar amount in stone was a risky proposition when inventing technology. But SwRI, like Ball, had wanted to get into deep space and had been willing to take the risk.

McLelland and his colleagues quickly realized how expensive their big leap would become. The same murky requirements frustrating Ball engineers translated into rework and engineering demands far beyond what the small team in San Antonio had anticipated. SwRI was soon months behind schedule and feeling enormous pressure. Deep Impact could evolve only so far without its brains. Everything connected to the spacecraft computers. A plethora of software tying Deep Impact's diverse systems together hinged on their delivery.

By early 2003, Deep Impact was in critical condition. The telescope mirror, the propulsion tubes, the gimbal, the gyroscopes, and the computers were the critical ills; but these issues were merely the most serious of many maladies the mission suffered.

After the cancellation review, Ed Weiler had sent the mission back to work with a bit more money—Deep Impact had $286 million to spend. But Weiler could still end it all at a moment's notice.

Chapter 17

Reprieves

John McNamee inherited Deep Impact from Brian Muirhead in late 2002. As the JPL manager learned his new assignment, he felt as if he had stumbled upon a highway accident. He didn't know what had happened, but it didn't look good.

To Deep Impact's good fortune, McNamee happened by like a doctor on a deserted road. He was one of two key changes in leadership—the other happened at Ball—aimed at stabilizing Deep Impact.

McNamee was one of the lab's few top-tier space managers with a background approaching Muirhead's in unconventionality. He had grown up in Florida and earned a business degree at the University of Florida in 1975. For ten years, he had been a financial analyst for the Southern Pacific Railroad and then a construction superintendent. He ended up in Austin, Texas, building homes.

But space had been a lingering fascination. His mom had let him stay home from school on his ninth birthday to watch the drama of John Glenn's orbital cruise unfold. More than 20 years later, after days spent managing framers and plumbers and drywallers, he read James Michener's *Space* and became intrigued with celestial mechanics, which involves calculating the motions of other worlds and the man-made creations aspiring to visit them.

McNamee stopped by the aerospace engineering school at the University of Texas and found it was among the world's very best at the discipline. He enrolled as an undergraduate while still working, did well, and went straight to a master's degree. He did well again, got into the PhD program in aerospace engineering, and earned his doctorate in 1988.

At JPL, McNamee worked on the Magellan Venus program, then climbed the project-management ranks, his experiences on low-tech construction sites dovetailing with his high-tech engineering expertise. A turn on Mars Pathfinder led to work on something called Mars '98. McNamee became project manager for the Mars Climate Orbiter and the Mars Polar Lander. When both failed in 1999, he had faced the press and public twice in the span of a few months. One crushing failure could end a project manager's career; somehow, McNamee had survived two. He brought to Deep Impact a conviction that no matter what the mission's price tag or what lip service was paid to accepting risk in return for lower costs, NASA, Congress and the public had no appetite for failure.

He recognized the challenge of having a deep-space neophyte as prime contractor.

"I don't think any of us really understood the overhead associated with bringing somebody into the world of planetary business. Ball was involved in kind of off-the-shelf spacecraft, Earth orbiters, a lot of good instrument development and the like and they're very good in that world," McNamee said later. "This was the first time when kind of a different language was spoken, and there were different expectations. A failure in this world is painfully visible to everybody. An Earth orbiter can fail, a military satellite can fail, and yeah, there's some news about it. But it's not the end of the world like in planetary."

McNamee arrived on a project in which technical difficulties were mounting, meetings were devolving into shouting matches, and NASA was losing faith. He set to work quickly.

First, McNamee spent more time at Ball than his predecessor. Muirhead was a single father and limited to visiting Boulder once or twice a month. McNamee became a frequent flyer and a consistent on-site presence at Ball, which he considered instrumental in getting the program back on track.

Second, he pried open communications between JPL and Ball management. McNamee believed, as had his predecessor Muirhead, that Ball management above John Marriott was "completely focused on the bottom line and not so much on mission success." McNamee turned Muirhead's one-big-team approach into something more arm's-length, where proposed changes would go through a formal process, translating added work into contract adjustments and more money to Ball.

Third, he delayed launch a year.

McNamee was amazed he could do such a thing. Planetary missions were slaves to the orbital mechanics he knew so well. Unlike an Earth orbiter, which could be launched whenever weather permitted, missions to planets or bodies that orbited the sun—comets, say—must depart at a certain time. Mars missions only happened once every 18 months when the planets were properly aligned. The comet Tempel 1 was less forgiving yet.

The Earth's position at Tempel 1's closest approach in mid-2005 suggested a January 2005 launch. But buried in the "announcement of opportunity"

NASA released for the 1998 round of Discovery proposals was a stipulation that missions be launched *"before the end of September 2004."*

Proposing to launch a few months later could have tipped the scales enough to lose out to a competing mission. Bill Blume, the JPL engineer charged with figuring out when Deep Impact should launch, came up with the idea of having the spacecraft orbit the sun for a year before heading off to Tempel 1. That way, Deep Impact could launch in late 2003 or early 2004, well in advance of its July 2005 encounter with the comet.

Mission leaders generally want their spacecraft to spend as little time in space as necessary, both to spare equipment undue punishment and to give engineers more time with their hardware on the ground. NASA proposal reviewers wondered why Deep Impact would dawdle about the sun for so long.

Mike A'Hearn explained it would give the team time to calibrate instruments and learn how to fly the spacecraft. It was true. But the Discovery Program launch deadline had been the ultimate reason.

Now, more than four years later, Blume sat in a meeting with McNamee and others trying to solve the mission's schedule dilemma. Almost in passing, Blume said, "You know, we could launch a year later. We don't have to do this whole loop."

McNamee blinked. He had assumed the year-long orbit was for an Earth gravity assist. JPL mission designers routinely harnessed the heft of planets to slingshot spacecraft to higher speeds and new trajectories without having to burn much fuel. Fuel being scarce on a spacecraft, you didn't just erase a gravity assist.

"We don't?"

It was as if they had inadvertently tripped a hidden switch and opened a secret passage to daylight. McNamee would only have to convince his own JPL bosses and then NASA headquarters to pay for the extra time on the ground.

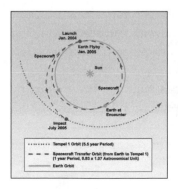
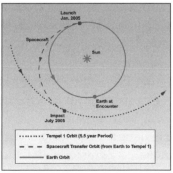

Deep Impact's mission plan, before and after the one-year launch delay

They concurred with the delay, and Ed Weiler agreed to spend for an extra year of engineering. In February 2003, he bumped the mission budget up to $299 million, the Discovery Program's absolute maximum.

THE EXTRA TIME WAS A GODSEND. But John Marriott, the Ball project manager, viewed it much less as a reprieve than as a deferred sentence. Graphs generated from his spreadsheets showed lines and columns with ground-truth "actuals" consistently outstripping the "planned" numbers against which NASA expected the mission to perform. Too many people were burning through too many hours. Deep Impact was spending too much money, too soon.

Changing course was like steering a freighter. Deep Impact was the biggest project Ball had ever attempted. Marriott had about 250 people at Ball alone, then another 50 at JPL, not to mention the subcontractors. He needed help.

Marriott was a gruff, honest man. He could be tactful, but often spared the niceties. He had spent years as a quality control inspector. Quality control was about finding fault. Rather than asking, "What have they done right," a quality control inspector asked, "What have they screwed up?"

But Marriott was more nuanced than many at Ball recognized. He had a keen nose for talent and fostered careers. He reached out to employees' families, who sacrificed as husbands and wives worked through dinners and weekends. He helped save marriages and, in at least one case, enabled one.

It happened not long after he had taken over Deep Impact. Before the move, he had been working on proposals. His office had been nearby that of a young Ball engineer named Marcy. They had gotten to know each other and talked about hanging chads, court decisions, and other controversies surrounding the 2000 presidential election. Then Marriott moved to the Tech Tower. When the announcement had been made that he would lead Ball's biggest project, he told Marcy he hoped it wouldn't affect their friendship.

"When you go home at night, there's nobody but you," Marriott later explained. "You don't get to go have beer on Fridays with the other guys. You don't get invited out to lunches and stuff like that. The separation is very large. It's hard. You're on your own."

One day, when Marriott was back in Marcy's building, he stopped by her office to say hello. She swiveled around in her chair and looked up at him. Rather than say hello, she burst into tears.

Marriott's fight-or-flight reflexes kicked in, but he resisted. She talked; he listened. She had been dating a guy for a year and a half. She had told him she wanted to get married. He had said he wasn't ready. So she had just dumped him.

Marriott could have offered some empty words of support, told her to take care of herself, and gone on with his day. She wasn't even working for him, and he had never met the guy. But instead, he took her out for burgers at the West End Tavern in downtown Boulder and listened. Then he explained the male's standard view of marriage and why men sometimes find the idea scary. He suggested she read a book about men and marriage, *What Women Want—What Men Want*.

Marcy and her boyfriend were back together in a week. She married him after all. Marriott was at the wedding. The guy was Lew Kendall, an attitude control engineer working for Marriott on Deep Impact.

Kendall knew that Marriott had saved his relationship with Marcy. He was someone who had as much reason as anybody to like John Marriott. But he couldn't bring himself to do it. He only knew Marriott as a project manager. The "You're gonna do what I say" Marriott. The flaw finder. The game face.

So Kendall considered the Ball project manager as someone "very interested in making the decisions to make himself look good." Others characterized Marriott as "a bull in a china shop," "iron-fisted," and "large and in charge."

Marriott was aware of this. He knew it wouldn't change, couldn't change, really, because he would be making tougher and tougher decisions—decisions that upended people's careers—as the project went on. He was also aware of what Muirhead had recognized: if Deep Impact was to ever get to the launch pad, the men and women on the program would have to put in an extraordinary effort as a team. That took nurturing, coaxing, and relationship-building. There was no way he could do it. Marriot had to play bad cop. He needed a good cop.

Monte Henderson was a good cop, the kind of guy who engaged security guards in conversation and remembered the details. He had been the software development lead for Marriott on the Spitzer Space Telescope, and Marriott could trust him. But in the summer of 2000, Henderson had left Ball. At that time, the dot-com bubble was fully inflated. An electronics manufacturer had made Henderson an offer. The company was within walking distance of his home in Louisville, a mining town turned bedroom community a few miles southeast of Boulder. Like many who abandoned established companies for promises of stock-option wealth, Henderson took the job.

Marriott, who also lived in Louisville, ran into him here and there. One afternoon at the hardware store, he asked Henderson the same question he always did: "Are you ready to come back to work yet?"

Henderson was not yet 40 and had a 3-year-old daughter and a pregnant wife. He had been away from Ball for a year. The dot-com bubble had burst. Yes, he said. I'm ready to come back to work.

Marriott put him charge of Deep Impact's instrument electronics, then promoted him to deputy project manager. Marriott would handle the "up and out"—Ball Aerospace's top management being "up," JPL and NASA Headquarters being the "out." Henderson would manage the day-to-day affairs within Ball.

Henderson monitored 17 teams, each doing something different, many working in areas in which he had scant expertise. But his job wasn't to understand the details. His job was to manage the teams. He had to sense when a team was cruising or in free fall. Many were in free fall.

"There are so many things that are not going to plan at a given time that you think, How can any of this actually come together and work?" Henderson said. "And it's a common feeling on every program. There's always those periods when you think, I can't turn around without some other turd hitting the punch bowl. And you're just . . . what the hell do you do?"

Disasters beyond his control contributed to his worries. Contour, the Discovery mission that was to zoom past three comets, had launched without a hitch and orbited Earth in perfect health for six weeks. Then on August 15, 2002, a solid-rocket motor built into the spacecraft was fired to send it into deep space. Contour was never heard from again. A NASA review board supposed heat from the solid-rocket motor may have melted the spacecraft's antenna or other key components. Or perhaps the spacecraft smashed into space junk, or simply began to tumble. No one would ever know. The board faulted the Contour team for poor analysis, weak systems engineering, and inadequate review and oversight.[1]

Joe Veverka, the Cornell University astronomer and Deep Impact science-team member, had led the Contour mission.

"The kind of stuff we're trying to do is challenging and exciting, and it's so challenging that occasionally it's not going to work," Veverka said. "So if you're not ready to accept that, you shouldn't be doing this."

Contour added to the pressure on Deep Impact, as did another, much more tragic disaster. The space shuttle Columbia, its wing wounded by a piece of fuel tank foam, disintegrated during reentry on February 1, 2003.

Not a few on the Deep Impact team had known the astronaut Kalpana Chawla, who had earned her aerospace engineering PhD while at the University of Colorado's Laboratory for Atmospheric and Space Physics, the erstwhile Rocket Project from which Ball itself had grown. In addition to the grief and the shadow of uncertainty the disaster cast across the American space program, it brought change to Deep Impact. NASA demanded more reporting and pressed Henderson and company to spend more on lowering the odds of another failed NASA mission.

Safety monitoring doubled at Ball and JPL. NASA Headquarters personnel combed through engineers' PowerPoint presentations for vague terms such as "some," "significant," and "nearly." The Columbia Accident Review Board had pointed to the inexactitude of "engineering by Viewgraphs" as having contributed to the shuttle disaster.[2]

Still, the basics of Henderson's job remained unchanged. He communicated. He talked to team leads—people in charge of propulsion, electrical, attitude control, the spacecraft body, software, testing, whomever. And he practiced an open-door policy. At first, nobody came, being accustomed to Marriott's command-and-control style. Then there was a trickle, and then the deluge—40 or 50 impromptu visits a day, interrupted only by meetings and phone calls and visits to shop floors and clean rooms. Lines would form outside his office. Marilyn Morris, an experienced administrative assistant working for both Henderson and Marriott, had never seen anything like it. "By the second week, people were just streaming in constantly, and he had no time whatsoever to do any practical work during working hours," she remembered.

So Henderson did his work at night—synthesizing what he had learned, prioritizing, scheduling, preparing for the weekly conference call with NASA, and, later, weekly meetings with Ball Aerospace CEO David Taylor, who took an increasingly keen interest in the project as it sank deeper in the red.

Henderson tracked various teams' progress with help from off-the-shelf project management software. The walls and cubicles of the Fisher Complex cube farm, where the entire Deep Impact team had moved, became coated with four-foot by six-foot bar chart schedules for 17 teams. An assistant to Henderson spent his days updating them, printing new versions on a plotter, and hanging them over the old ones until the thickened layers crinkled in the breezes of passersby.

Teams worked hard to avoid falling behind schedule, which could put them on the critical path. Avoiding the critical path was often a game of chicken, with subsystem leads keeping tabs on the other teams' charts and shaving a day or two off their own schedules just to avoid the most intense management attention. Henderson, having been a subsystem lead himself, made a habit of closely watching all the teams within shouting distance of the critical path.

•

THE GROUP BUILDING THE IMPACTOR had earned particular scrutiny from Marriott and Henderson. The team was behind. The real issue, though, was that Ball management came to see the product of the team's work as too...smart.

In other circumstances, David Wilson might have considered this a compliment. Wilson, the self-described mercenary who had led the team creating Deep Impact's electrical system, was atop the impactor effort. He had accepted the role reluctantly. It had come about during what became known as the "May Massacre" of 2003, when fully 30 percent of the project's staff was cut for lack of money.

Wilson had been leading the power-system team when a Deep Impact manager dropped by. He suggested Wilson consider taking over the smaller spacecraft.

"No," Wilson answered. "I just built this power team from nothing, I was way behind, I'm on track, things are coming together. I'm not interested."

The next day someone else stopped by and suggested the same thing. Wilson declined again.

On the third day, Marriott called Wilson into a conference room. Several of Marriott's deputies and other project managers were already there. Wilson sat down. It was quiet for a moment. Then Marriott turned to him.

"Do you care about the rest of your team?" Marriott asked. "Do you care about the people you work with?"

Wilson exhaled.

"You'll do this job then."

Wilson was new to managing spacecraft development. He took over a team with morale problems. Deep Impact was two spacecraft. If Tempel 1 was to take a bullet in the name of science, both had to function. The smaller craft—the bullet, the deliverer of pyrotechnics for which Ed Weiler and NASA agreed to pay on the order of $300 million—was at least an equal partner.

But the flyby was more complex, dealt with communications to Earth, had the fancy gimbal and the reaction wheels and the big telescope and the solar panels and the nifty spectrometer. Most of all, the flyby had a future. Never mind that the impactor was the whole point of the mission. It was a cudgel, a flying anvil, a creature with the life span of a mayfly.

If shift work had to happen, the flyby got days, the impactor nights. Impactor engineers started to feel like the B team, perform like the B team. Wilson's job was to turn it around.

He began by considering his team as a glass half full.

"Success on a program like this is highly dependent on your leadership and whether people are being led or just being left to their own devices," he said.

The impactor spacecraft in testing

Alec Baldwin, left, guides the impactor spacecraft onto its mount. Three layers of Whipple shielding plus the double-layered aluminum honeycomb spacecraft body were designed to protect the impactor from comet-dust impacts.

"Because most of these people have tremendous depth, drive and capability. It's just whether they're being used well or not."

It was a perfect role for Wilson, who operated best, as Henderson put it, in an "us-versus-them, screw-the-man-we're-gonna-get-it-done" environment. Wilson cross-trained people on different parts of the spacecraft and refused to let problems with a particular subsystem stall progress. Work on the flyby spacecraft would get bogged down, Wilson said. On the impactor, "We would move. No matter what."

But Deep Impact was sinking deeper and deeper into the red. The obvious "descopes"—the industry term for hardware thrown off a spacecraft—were long gone. To Marriott, the mission needed to do something drastic. He called Wilson into his office. He explained the plan, and then gave Wilson some management advice. "Don't wear your emotions on your sleeve," he suggested.

Wilson went to Jeremy Stober, who had helped design the impactor spacecraft. Stober was now the lead systems engineer in the small craft's construction. Wilson began asking "crazy-ass" questions, as Wilson later described it, such as, "How much would it cost to just bolt all the boxes on and connect all the harnesses and put the shell on and say, 'We're done.'?"

In other words, to kill the impactor and send its lifeless corpse up with the flyby. Wilson was shielding him and the other 30-plus people working on the impactor from what Stober described as "management pain." Still, Stober recalled, "I mean, you're basically writing out your death sentence."

It would save perhaps $10 million dollars, Stober reported back to Wilson.

Wilson reported that to Marriott. Marriott quietly ran it by JPL. His argument was simple. We can save $10 million, which could keep Ed Weiler and NASA from shutting us down. If the flyby spacecraft were to hang onto the impactor longer—releasing it 12 hours before encounter rather than 24—we'd still hit the comet. The requirements were to hit the comet, Marriott reminded them, not to hit the comet with an intelligent homing device.

The science team was displeased. Remember the 1996 proposal, rejected to no small degree because of the dead impactor? Deep Impact existed as a space program *because it had a live impactor*. And you want to kill it? No way, A'Hearn insisted.

Even Jerry Chodil, the Ball Aerospace vice president overseeing Deep Impact and several other programs, questioned the idea. "It's like if you're building a car and you've got all the pieces coming together and you've got a cost overrun, the question then comes, 'Well, should we only put three wheels on it rather than four? Or should we leave off the steering wheel because it cost too much?'" he said. "No, you can't. Because the whole thing doesn't work. So the only time you can do these descopes is very, very early. You just don't go pull a tire off as it's coming off the assembly line and say, 'Well, that saved us $100.'"

Or $10 million.

So the impactor came to life again. Weeks or months later Wilson would call Stober to his Fisher Complex cubicle and ask more of his crazy-ass questions. Finally, after the third or fourth time, Stober "just kind of ignored it," he said.

Marriott would have gone through with dumbing down the impactor. But mostly he was sending NASA Headquarters a message. We have cut every last bit of fat from Deep Impact. If you want to hit a comet to see what's inside, you're going to have to either kill us outright or cut us some slack.

•

NASA Headquarters did not accept this quietly. Ed Weiler had made a deal with Deep Impact, and the comet mission was failing to deliver. He and his deputies were not above hollering on conference calls. How are you going to get this under control? Look at all the other missions we've cancelled! Why shouldn't we cancel you? In late 2003, the Deep Impact team landed back in a Washington, D.C. conference room. To this cancellation review, Marriott had the pleasure of flying in the Ball corporate jet. NASA had requested the presence of David Taylor. Marriott briefed the Ball Aerospace CEO during the flight.

Taylor had it relatively easy at NASA Headquarters. From him, Weiler just wanted to know whether Ball Aerospace was committed—*really committed*—to the mission's success.

Yes, Taylor said, he was really committed. In fact, as a demonstration of his and his company's devotion to Deep Impact, Ball would buy back from NASA $7 million in Deep Impact ground-support equipment, an amount roughly equivalent to the "fee" or profit, built into the mission. Aerospace firms usually pay a dime on the dollar for such hardware.

This act was less out of heartfelt generosity than a backchannel understanding that the sacrifice was the price of keeping the mission alive. For Taylor's company to undertake a project of such magnitude without profit was a major financial hit. But Taylor wanted this leap at deep space as much as anyone. He believed in the team and he considered Deep Impact vital to his company's future with NASA. He had insisted to his own Ball Corporation bosses that the sacrifice was worth it.

The meeting had an entirely different tone than the first. Later, it would be remembered not as a cancellation review, but as a "continuation review." NASA chipped in more money, boosting the total to $318 million, violating its own Discovery Program cost cap in the process. Why? It was clear to all involved: Deep Impact was turning a corner. The extra money was to fix Deep Impact's biggest problem and greatest remaining risk: the computers.

Chapter 18

The Spacecraft is the Computer

Consider again the Orbiting Solar Observatory, Ball's first spacecraft. It achieved something extraordinary—collecting and sending home a trove of new information about the astronomical body responsible for life on Earth. The satellite managed to point at the sun, turn on instruments, turn off instruments, collect data from instruments, store the data on magnetic tape and transmit it—all despite being able to obey just 10 commands.

Transistors existed in OSO's day, but they hadn't been miniaturized and compressed into computer chips containing millions of semiconducting on-off switches—microprocessors. As computers shrank, the amount of processing power engineers could load on their spacecraft took off.

What did processing power do for spacecraft? Raw speed translated into an ability gather exponentially more data, enriching the scientific returns from a given spacecraft. But the biggest difference was in sophistication. Computers changed the nature of American spacecraft as profoundly as they changed American life. Rather than 10 commands, Deep Impact could respond to thousands and issue its own. The spacecraft could take its own temperature in dozens of places, turn its own power up or down, understand where it was aimed in space and rotate itself. It could talk with impactor and Earth at the same time. It could compress, store and send images and spectral data captured on four different light-sensitive computer chips. It could sense if it was sick, diagnose its ills, and in many cases remedy them. It could split itself in half at the right time; it could steer itself into a comet and watch the carnage with another part of itself. It all required computation. Deep Impact, like all modern spacecraft, was a flying computer.

Alas, once spacecraft became computers, they became subject to the laws of computer systems. A NASA report published in 2000 took a broad look at U.S. software development. It found that, of 8,000 software projects across various industries, 31 percent were canceled during development and more than half suffered some sort of major delay or cost overrun. Just one in six projects were "fully successful," and only one in 12 "major" projects succeeded. Even

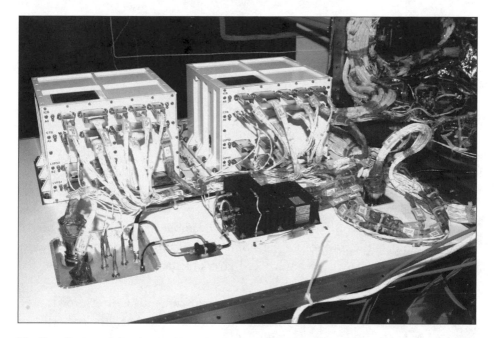

The Deep Impact flyby spacecraft's dual main computers

those were riddled with errors—five to 10 mistakes per 1,000 lines of code in commercial products.[1]

That was just the software. Terrestrial projects ran on proven hardware mass-produced by IBM, Apple and Dell. The Deep Impact team was inventing not only their software, but also the computers running it. The RAD750-based computer was a unique machine.

If a problem cropped up, someone had to determine whether it was a hardware or software quirk. Solving a hardware problem often exposed a software problem, which, once addressed, exposed another hardware problem.

Complicating matters was the division of labor inherent in the "mind" of Deep Impact. The computers on both spacecraft compartmentalized their functions. The Remote Interface Unit dealt with such things as the spacecraft's physical coordination (attitude control, propulsion, the Starsys mechanisms to spring the solar panels open) and temperature control. The craft's more cerebral telecommunications, instrument control, and general operations happened in the Spacecraft Control Unit. In a box that could hold a basketball, this cerebrum also housed the RAD750 microprocessor and the 512-megabyte, cosmic ray-resistant flash memory. This nonvolatile memory served as a faster and more reliable version of OSO's tape recorders, storing images and data from instruments as well as the software and commands to run the spacecraft. Connecting the two brain regions—and in the case of the flyby spacecraft, the redundant cerebrums themselves—was a spinal cord called a military-standard 1553 serial data bus.

Amy Walsh, a young engineer on the avionics team, a.k.a. the Command & Data Handling team, was at the center of the effort to get the computers working. Walsh was in her early thirties, tall, thin, with straight brown hair. After work, if the weather was reasonable, she donned a leather Fieldsheer jacket with blue sleeves worn brown at the elbows and a bright-yellow helmet adorned with what appeared to be either an eagle or a pair of cobras. She straddled her 1988 Honda motorcycle heading up U.S. 36 north out of Boulder and past the town of Lyons, home of the Outlaw Saloon, a beef-jerky stand and the quarry whose repurposed layers formed many of Boulder's red sandstone structures.

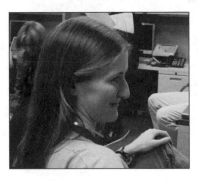

Amy Walsh

She turned right up County road 71N, where Angus cattle, llamas, horses and deer nibbled parched grasses amid ponderosa pines and rocky outcroppings. Up Lonestar Road and Stage Trail she motored, out of Boulder County into Larimer, and then veered onto a dirt road with no street sign, where she stood and leaned into her handlebars and powered up the homestretch.

There, 27 miles from Boulder, she lived in a small house overlooking a valley. An outbuilding, guarded by an imposing wolf-huskie mutt, housed a dozen motorcycles, including the dirt bike Walsh raced. Other bikes belonged to her boyfriend, Steve Ling, nicknamed "Flash." Still other bikes belonged to Flash's clients. Flash repaired motorcycles.

They had met at the Boulder motorcycle shop where he had worked between arriving from Salisbury, England, and launching his own business, Sunrise Offroad Motorcycle Service & Prep. He picked up motorcycles in a truck with custom plates reading "Flash-2," fixed them up in his shop, and drove them back down.

From the time Walsh was eight years old, she had split time between her dad's home in Carpenteria, California, and her mom's in Steamboat Springs, Colorado. Every year, she uprooted and resettled with the opposite parent. In neither place did she live the resort life. Her mom cleaned hotel rooms and did other odd jobs; her dad on different occasions managed a Thrifty drug store, drove trucks, did construction, and, eventually, worked with computers.

The daughter came upon computers indirectly. She thought she might be an architect, then decided her creativity was "more functional and less artistic."

"It has to do a certain thing. It doesn't have to look a certain way," she explained.

She excelled at UCLA and hired into the Jet Propulsion Laboratory in 1993, where she worked on telecommunications for the Cassini Saturn orbiter and next-generation deep-space transponders. She came to Ball in 2000 and worked on Deep Impact. Her first job was to design the mother of all boards for the spacecraft computer, which was the first computer board she'd ever designed. Southwest Research Institute later built something close to it.

Her creation, the command & telemetry board, was an information hub directly or indirectly touching nearly everything on the dual spacecraft. Walsh learned a lot about Deep Impact's systems. Because she knew so much, she could help solve all sorts of puzzles. Walsh liked puzzles. In fact, she was driven by puzzles. Deep Impact wasn't about some higher purpose, such as, "Oh, these scientists really need their data,' or 'My management really wants me to do this,' or 'Ball needs me to work hard,'" she said. "It's all about, 'What the hell's going on there? I've got to figure out what it is and fix it.'"

The puzzles seemed less imposing at the beginning. Early in the spacecraft's development, she hiked among the Flatirons over long lunches with colleagues. Then her father's ill health forced her to California, where she worked for a solid month from his hospital room. As he slept, she sat on a rollaway cot, the glow of her laptop screen like a candle in a cave. She dialed into Ball's servers using a line poached from the room's phone, sifting through data or answering e-mailed questions from Southwest Research and her Ball colleagues.

The hikes became a pleasant memory. Her work days got longer. There were many reasons. One was that Southwest Research had signed a contract to deliver entire computer boxes but could test only as far as the boards within. For one, they lacked the sophisticated testing software in place at Ball. SwRI would ship a machine to Boulder, where Ball would test it. Something would go wrong. Ball would send it right back to SwRI. For two years, a steady stream of computers commuted between Texas and Colorado.

Sometimes they traveled FedEx Custom Critical, where a minivan shows up at the door and drives a single box from San Antonio to Boulder. Often it was easiest to buy a first-class airline ticket for a spacecraft brain and send an engineer along next to it. On one trip, Buddy Walls, the Southwest Research engineer in charge of the main spacecraft computers, ordered his traveling companion a drink.

"Pardon me, sir?" the flight attendant asked, glancing at the large gray plastic box the passenger seemed to be referring to.

"It has a ticket," Walls clarified, still smiling. "And it wants a beer."

It got a beer.

Walls received e-mails from Walsh at 2 a.m. *For the love of God, the woman doesn't sleep*, he thought as he read them the next morning.

Walsh's workweeks stretched to 80 hours and beyond, with no days off. She wasn't sure how it had come to this. People asked questions about how something worked, and the answers were complicated. "So rather than saying, 'Look here, look here and look there,' it's sometimes easier to solve it yourself," she explained. She developed a reputation: exceptionally thorough, exceptionally bright. "She had a little chip on her shoulder but a deserved one," as a colleague put it. "You really couldn't question her. You could ask her stuff. But question her and she's got like five different things to support her point of view. She was right, usually."

Her boyfriend called at eight, nine, ten at night to see where she was. Still at work, trying to solve this or that puzzle, she would tell him. On the other end of the line was a heavy silence.

"I just felt Amy was wiping their asses for them," Flash said.

It got so that she and Flash would "hot bunk," with her rolling into bed as he rolled out. An occasional dirt-bike ride provided her only escape. At one point they met at the trail at eight in the morning. She had come straight from work.

The term "Deep Impact widow" became part of the mission's parlance. Flash, the burly motorcycle repairman with the British accent, was foremost among them.

Tom Golden had no Deep Impact widow at home. He was a software guy, in his early twenties, from Fort Worth, Texas. There he had coded cell-phone equipment for Motorola since graduating from the University of Texas at Austin in 2000. He had decided he wanted to do something cool and live someplace nice. Since arriving in Boulder in early 2003, he had experienced primarily the interior of his cubicle.

It was gray and unadorned—just his laptop, the server containing various versions of Deep Impact's spacecraft software, a 21-inch CRT monitor consuming half his desk. Only the color photo on the Ball Aerospace calendar (employee nature shots) interrupted its stark utility.

He saw software as the real brains of a spacecraft. "All the hardware's dumb," he said. Software, absent on OSO, was pervasive on Deep Impact. "It's been growing over time, and now it's this huge behemoth beast."

The beast had been largely built by the time he arrived. Rather than using off-the-shelf code as had been planned in the original proposal, Deep Impact's software was brand new, based on a Ball research and development effort called Advanced Space Electronics ("Aspen") Flight Software.

Ball's idea was to create a next-generation spacecraft operating system—a Microsoft Windows or Apple OS for computers that flew. Then, rather than rebuilding the next spacecraft's software from the ground up, one could port over the Deep Impact core and customize at the margins. It would be modular throughout, with battle-tested applications ready for plugging in like so many virtual Legos. The software would handle the basic operations of a spacecraft computer in new ways, and allow a computer to control a spacecraft's functions in an extremely fine-grained manner. If Ball's existing spacecraft computers played in whole or half notes, Aspen could handle sixteenths. It had been, in part, the processing demands of the new software that had led to the need for a new, RAD750-based computer.

The Aspen software remained unfinished when Deep Impact began. The comet mission was to share the burden of completing it with two other missions. But those two missions fizzled out, so the team creating the fancy new software was absorbed by Deep Impact, and Deep Impact, with its Discovery Program budget, was suddenly developing another entirely new technology. It was a monumental undertaking. About 320,000 lines of code were to ride the flyby spacecraft alone. *Moby Dick*, for comparison, is 215,000 words long. Complicating matters was the software team's being split between JPL and Ball, with JPL handling the fault-protection system. This software, too, was the spawn of a research program, albeit at JPL.

Golden was one of a dozen people on Ball's Deep Impact software team, an expert in something called embedded software. Such code, operating invisibly in a dizzying array of devices, is single-minded in its pursuit of making computers run better, usually under the flag of "firmware."

Like Walsh, Golden was pulled into problems with Deep Impact's spacecraft systems that his "firmware" touched. He learned enough about the instruments to become ensnared in one of the Live for the Moment strategy's most visible consequences: imaging performance.

At encounter, Deep Impact's flyby instruments became the equivalent of two movie cameras shooting at 16 frames per second and blasting the data into the spacecraft computer like a pair of water cannons. During testing, one image racing in from the instrument electronics was stepping on the next. Golden was asked to wring out better imaging performance.

We could use another 10 percent, his boss told him. And we need it in two weeks.

Golden fueled himself with peanut M&Ms and Cokes and Mountain Dews from the mini-fridge in the next cube, nicknamed Boyd Mart in honor of its inhabitant. He got to know every data pathway: where it came in, how it came in, where it was stored, how it moved from one piece of software to the next. He tweaked, refined, stretched and patched—all, he hoped, without breaking some other bit of the C++ code running Deep Impact.

Golden picked and chose fixes to win a half-percent here, a percent there. He got to what he thought was 10 percent and moved his updated software into the formal testing process. Running a short test took six hours; the long ones took 24 hours.

Finally, he was done.

Then the science team decided they could use a few more images, and new mission-sequence products—scripts from which the spacecraft acts out its role in space—came in. "You're like, 'OK! We meet the performance! Great! My job is done!'" Golden said. "And then a week later they're like, 'Here's all-new products with 20 percent more images.' And you run it and it breaks."

Golden started all over again.

"It was extremely frustrating," he said.

The term "test bench" evoked experimental park furniture. The Deep Impact mission's test benches were, rather, a formidable combination of computing technologies. The mission had four test benches, two at JPL, two at Ball. If modern spacecraft were flying computers, the test benches were flightless spacecraft. The test benches ran through the motions of launch, cruise, and impact, ferreting out problems such as the imaging-speed issues Golden helped fix.

The test benches varied in their specifics, but all included at least one clone of a flyby spacecraft computer, which was connected to custom-made blade servers filling a rack the size of a refrigerator. The rack, in turn, was connected to more familiar-looking computers with keyboards and monitors.

First one test bench, then, starting in mid-2004, a second lived on the eighth floor of Building 264, home of JPL's Deep Impact team. One bench had two spacecraft computers, simulating the flyby spacecraft; the other had three, simulating the main computers for both flyby and impactor. Connected to each box were two dozen cables, each snaking to the neighboring computer rack, which in turn was connected to computers simulating ground systems. Taped to the apparatus holding the clone spacecraft computers were two signs. One read, "Caution: Deep Impact EM Hardware/Do Not Disturb." The other read, "Caution: Stress Pheromone Collection Apparatus In Use."

Ball had done simulations before, but nothing approaching the complexity of the Deep Impact test benches. All spacecraft need to be well-tested. But Deep Impact, unlike spacecraft Ball had previously built, faced what JPL engineers called a *critical sequence.*

A critical sequence is something a spacecraft must do on its own or die. A Mars landing is a critical sequence. A Saturn orbit insertion is a critical sequence. Deep Impact's comet encounter would be a critical sequence—or, more accurately, two critical sequences.

Critical sequences happened far enough from Earth that light-travel time took human intervention off the table. If something went wrong, the spacecraft had to figure it out and recover on its own. Hence the cross-strapped computers and the ornate fault-protection system JPL was creating for Deep Impact.

Earth orbiters knew no critical sequences. If something bad happened— things got too toasty or a computer routine crashed—the spacecraft turned its solar panels to the sun, shut itself down, and called home for help. If Deep Impact did such a thing to close to encounter, Tempel 1 would race by before advice from Earth arrived.

Spared critical sequences, Ball's orbiters had little need to test every possible operational wrinkle on the ground. So the idea of high-fidelity test benches was new to Ball. And it was new to Sean Esslinger, whose job was to help create them.

Esslinger, in his early thirties, looked like a former high-school football lineman. The son of a general contractor and a homemaker in Louisville, he had, rather, spent afternoons during high school and then at the Colorado School of Mines rebuilding his 1974 Jeep Cherokee or earning enough money to buy parts for it. His habit of rolling the Jeep and battering into other off-road enthusiasts made it a long-term avocation. The Jeep was still in his garage when he joined Ball in 2001.

The two Deep Impact spacecraft (the impactor is in the foreground with its hexagonal top cover removed) in testing. Dynamic Spacecraft Simulator software helped engineers fool the spacecraft into believing they were already flying through deep space.

On Deep Impact, his charge was to build something called the Dynamic Spacecraft Simulation.

"We make the spacecraft think it's flying," he explained.

First, he had to make the spacecraft's test-bench computer—really the equivalent of a brain in a jar—believe it was a fully-formed spacecraft.

He and a colleague or two, depending on the month, had to learn how every system on the spacecraft interacted with the computer and then imitate those systems in their software. The test benches existed well before the actual flyby and impactor spacecraft had been assembled.

When the spacecraft did come together, Esslinger and colleagues adapted the test-bench system for the spacecraft itself, tapping into the spacecraft's 1553-bus spinal cord. If Deep Impact's computer commanded "turn on instrument heaters," the software would obey and dissimulate. "The heaters are on," the software might say, and even send over a few imaginary numbers feigning rising temperatures. The software could mislead the spacecraft computers into believing that solar panels were delivering a certain amount of power, that reaction wheels were spinning at a certain rate, that antennas were sending and receiving instructions, and that thrusters were blasting noxious exhaust into a computer lab in Boulder or Pasadena.

With his simulated universe, Esslinger convinced Deep Impact's disembodied brain it was already flying in space. He fed gyroscope readings scrubbed of all Earthly influence; he supplied the coordinates of stars from faked star trackers, simulated the orbits and gravity of the nine planets (Pluto had yet to be demoted), feigned solar pressure on the spacecraft, and made up the comet's location, size and motion.

But Esslinger, if an accomplished software engineer, was an imperfect god. There were quirks in various simulations, sometimes due to his own miscalculations, other times because something changed with the spacecraft itself and someone forgot to mention it to him. As a result, problems on a test bench had yet another possible source. They could originate in spacecraft hardware, its software, or they could stem from the simulated hardware and software Esslinger's code mimicked.

With time, he ironed many problems out. But money was short, and the test benches never received the attention they deserved. This would come back to haunt engineers at JPL and Ball later in the mission, when the health of the flightless spacecraft would determine the fate of the real ones. First, though, Deep Impact would have to prove its physical mettle.

Chapter 19

Mating

While Joe Galamback and Alec Baldwin bolted, glued, and plugged more and more hardware onto their planks of honeycombed aluminum, Deep Impact's largest team, the Integration and Test team, or simply "I&T," tested and tested. Their work took many forms, from checking on software routines to just beating on the hardware, sometimes literally.

Even as Lorna Hess-Frey and colleagues had mulled over divergent spacecraft designs, test engineers were plotting their long campaign to assess whatever product might emerge. Their work also involved building an arsenal of specialized hardware to hold, lift and move their evolving spacecraft, as well as the test-bench computers and software making the spacecraft feel like they were flying.

Then the brain arrived. Or a version of the brain—a prototype at first, which stayed in place months longer than expected because the flight computers were so long delayed. Then finally, in August 2003, the space-ready computers rolled in. And then they rolled back out again. Computers that Southwest Research said were ready stumbled instead, and Galamback and Baldwin unscrewed them and sent them on another first-class trip back to San Antonio. They then put the prototypes back on.

Finally, in early 2004, the spacecraft's flight computers stayed. Monte Henderson had flown to San Antonio to witness the final shock test. It involved a section of lead pipe roped up as a pendulum. The pipe was to crash into a metal plate holding an operating, flight-ready spacecraft computer worth many millions of dollars. This was to be repeated four times. The test had been meticulously planned, but Henderson couldn't quite get over how medieval it all seemed, especially considering the velvet-glove treatment the boxes had otherwise been getting.

"Why don't you just tie it to your bumper and drive the thing back to Boulder?" he quipped to Buddy Walls, the Southwest Research engineer.

"Hey, they're your requirements," Walls answered.

A technician cut the cord suspending the pipe. It fell like a sledgehammer on a manhole cover. Henderson winced.

With the arrival of the flight computers, the test team fell into a months-long push in which two shifts were the norm, and three-shift, 24/7 testing was not uncommon.

Nick Taylor ran the testing effort for Ball. Born in England and raised near Edinburgh, he had worked on the space shuttle arm and other space hardware in Canada. Taylor had spent a brief period working on an ill-fated Internet-in-the-sky effort at Motorola in Texas. Then he came to Ball in 2001, not long after his fortieth birthday.

For Deep Impact, he had test plans and backup test plans and backups to those. If a piece of hardware or software was late, they ran something else. No matter what happened, testing had to move ahead, he insisted. "The way to get the job done is to utilize the hardware on the floor," he said.

Until the spacecraft testing phase, engineers mainly found their own problems. Taylor's team, innocent of spacecraft hardware and software engineering, sought out the mistakes of others.

A tester's work evolved as the spacecraft took shape. A box would arrive, a "box" being a tester's term for a custom agglomeration of electronics worth more than a house in a good neighborhood. It might have been part of Deep Impact's telecommunication's system, say—a "tweeta," or, formally, traveling wave tube amplifier (TWTA), which was an amplifier powerful enough to make a spacecraft heard across millions of miles. Deep Impact's tweetas were the first Ball had dealt with. They didn't need tweetas screaming down from low-Earth orbit.

The tweetas had come from JPL. They were an engineering model and a flight spare left over from the Cassini Saturn orbiter. Both had been tested and parked for years (used as doorstops, maybe, Ball engineers mused) with no expectation of seeing space.

On the flyby mother ship, telecommunications boxes and electronics lived on the underside of the spacecraft's roof, right below the big high-gain antenna and the exquisite gimbal steering it. There were two transponders (in case one failed) dealing with high-speed X-band communications to and from Earth, as well as the amplifiers and other electronics. The flyby and impactor also carried antennas, amps and electronics for the S-band microwave communications between the two spacecraft once the impactor ventured off on its own.

Larry Murphy, Deep Impact's telecommunications lead, had worked the kinks out of his system before Nick Taylor's test people had a shot at it. The mission's telecom system thus had been tested at JPL years before and again of late by Ball's most experienced telecom man. Taylor's testers took no interest in pedigrees. They approached the telecom system as if it had been built by bonobos. They checked every connector plug and buzzed every pin, assessing myriad voltages. If something seemed amiss, they referred to cable drawings or engineering schematics, because a good test engineer, like a good doctor, suggests remedies.

When one tweeta passed, they repeated the process with the second tweeta. They did the same with each transponder, then with other boxes. They ran computer-driven tests to see whether the telecommunications system did what it was supposed to do. If not, testers filled out a form with their diagnosis and suggested solutions. Such sheets, called test anomaly worksheets, eventually numbered in the hundreds. The Deep Impact team retired them by fixing spacecraft software (or, more rarely, hardware), identifying a quirk with the model universe, or dismissing the problem as an error in the test routine itself (a tester instructing the spacecraft to tie its shoes before putting its shoes on, say).

The two spacecraft gradually became presentable. The flyby and impactor occupied part of the big Fisher Complex clean room. Those entering wore "bunny suits"—clean-room garb, white with the sky-blue Ball Aerospace logo between the shoulder blades—to preserve an atmosphere ten times cleaner than typical office air. The two spacecraft sat about 20 feet apart. The scene was one of organized chaos: voltmeters on folding tables, oscilloscopes on rolling carts, the impactor spacecraft with its top cover removed. The flyby, lacking its imposing, swarthy solar panels and big antenna through most of the testing regimen, seemed less significant, a Batman without cowl or cape. A bank of computers running versions of the test benches Sean Esslinger and others had conjured up connected to the spacecraft through a web of wires and cables.

The clean-room test bench adapted with each added system. When thrusters came aboard, the test bench no longer simulated the thruster electronics—the spacecraft computer could talk directly with the propulsion system. But then the computers controlling the thrusters had to be deceived into believing butterfly burps were happening. So the clean-room test bench's software sniffed for spacecraft commands to fire thrusters and responded with white-lie confirmations to the propulsion system. Other simulators flooded the spacecraft with made-up attitude and trajectory-change information, star locations, energy coming from the solar-panels, the gravity of Neptune, the warmth of the sun.

The test team was diverse in experience and temperament. Craig Long, who started out working on Deep Impact's propulsion and power systems, was extremely quiet, focused and tenacious. Faced with the unexpected, he divided and conquered, parsing and digging until he cornered even the most obscure problems, forcing them to reveal their nature, dooming them to resolution. A colleague described him as "one of the baddest-ass testing machines I've ever seen. There is no, like, getting hungry. There is no getting tired. There is no complaining about anything. It's like, dude is a freaking testing *machine.*"

The colleague was Jon Mah, who was as fiery as Long was reserved and, at 24, the youngest member of Ball's Deep Impact team when he started in 2002. Mah's parents had come from Korea in 1971 and earned technical PhDs after moving to Ohio, where Mah grew up. He was the product of a relatively recent development in the history of Ball Aerospace, having been hired straight from Purdue University's engineering school. He had had some industry experience, through a co-op program alternating him between studies in Indiana and work at Hughes in El Segundo, California, which became Boeing Satellite Systems in 2000.

Mah had worked on massive geosynchronous satellites (racing through space in an orbit 23,000 miles from Earth, they seem to hang motionless in the sky) for beaming satellite radio and television signals across continents—HS-601s and HS-702s, with solar panels 20 times the size of Deep Impact's. They were as close to one-size-fits-all as you can get in the space business, and Boeing cranked them out, with two dozen or so in varying stages of completion at a given time. Like Ball's Orbiting Solar Observatories or CU's biaxial pointing controls, payloads had to adapt to the bus, and not vice versa. To Mah, the contrast with the products of Ball Aerospace couldn't have been starker. Payloads here seemed to have the power to bend and shape spacecraft at will.

Mah worked with Long on the impactor, the more experienced man teaching Mah the ropes as the young engineer tested attitude control and thermal subsystems on the smaller craft. As the program went on and the dual spacecraft became electrically united, Mah took on telecom, thermal, and instrument tests for the combined system. He racked up 60-hour weeks and ended his long days with late-night hockey games and a few drinks with teammates. On a few hours sleep, he slogged into Nick Taylor's 7:30 a.m. test team meetings, long hair ponytailed or loose and wet after a rushed shower. On one occasion, his team broke out in applause when he came in.

Mah was of a different generation, one still at home in Boulder's many college bars. He held fast to the idea that, "If you're going to be a man at night, you've got to be a man in the morning." One morning, Monte Henderson took one look at his hung-over subordinate and said, "You're not going anywhere near that spacecraft today."

Mah, being a man in the morning, ignored him.

MOST SPACECRAFT ARE CONCEIVED and developed. Mating is less common.

Deep Impact mated, in public, on a Wednesday afternoon in April. Although the tryst had been arranged years earlier, how well the pair would connect on this spring day in 2004 remained uncertain. These were two very different beasts: one compact, aggressive and fatalistic; the other far larger, sensitive and shy, preferring to observe events at a distance.

The pairing was uncommon enough that it attracted a Discovery Channel documentary crew. The crew filmed from a corner of the room, far from the action, as if behind a blind on the Serengeti.

Despite the logic of placement, the flyby craft's hardware seemed haphazardly arranged: gold-wrapped telescopes at odd angles, the awkward main antenna, conical protrusions for reading the stars, the sleek solar panels. Spacecraft, at home in space, were misfits on Earth.

The flyby hung like a puppet whose master had gone to lunch, from a bridge crane spanning the width of the clean room over a concrete floor painted brown. Despite the crane's resemblance to beams of a highway overpass and the thickness of the various cables holding it, and despite every hook and bolt associated with the lift having been qualified and tested many times over, those who had dedicated four years of their lives to getting this technological wonder built found the scene unsettling. The dangling spacecraft was worth several times its weight in gold.

The impactor, the size of a washing machine but resembling nothing Sears sells, waited on a nearby stand. Mating involved using the crane to maneuver the flyby spacecraft over the impactor and then lowering it with care. The impactor's hexagonal upper shell would—or should—enter a like-shaped cavity at the bottom of the flyby spacecraft. The tightening of four explosive bolts and plugging together an umbilical link would seal the consummation.

Fourteen months later, the bolts were supposed to explode in perfect unison, divorcing the spacecraft as abruptly as Hollywood stars.

Before the spacecraft could move tens of millions of miles, it had to move a few feet. Galamback, Baldwin and a few colleagues wore bunny suits, plus gauzy caps and facemasks, rubber gloves, and white pant- and shoe-covers. Galamback served as "lift coordinator." He had been the last to leave the clean

The flyby spacecraft hangs over the impactor, preparing to "mate."

room the night before, around 11 p.m. He had looked back toward the two spacecraft, connected by data cables to computer terminals. He had reached for the light switch, then paused.

"It was such a huge milestone, and all of a sudden these thoughts start running through my head: 'Oh my God, we're not going to be able to lift the thing high enough on the crane to integrate it tomorrow. And not only am I going to have the customer and Ball management there, I'm going to have the damn Discovery Channel filming this,'" Galamback recalled.

The crew was filming for a Deep Impact documentary to run on the cable TV network. It was a high-profile addition to Deep Impact's formal education and public outreach effort, a common denominator of all major NASA missions. For Deep Impact, it included a Web site; a monthly electronic newsletter; educational materials for classroom use (including "make a comet and eat it" and "comet on a stick" activities); a program to help amateur astronomers collect information on the impact; the Send Your Name to a Comet program (yielding 650,000 names for burning onto a mini-CD mounted on the impactor spacecraft); even a "Comets" song in both standard and rap versions *("Comets are very cold/from the Oort cloud they rolled....")*.

For the Discovery Channel, engineers became actors, scribbling spacecraft designs on Plexiglas, examining with feigned worry a cracked telescope mirror that had been replaced two years earlier, holding informal meetings on the Fisher Complex roof, all for the benefit of cameras. But the mating wasn't scripted. If something went wrong, a television crew would record the fiasco in embarrassing detail.

Henderson had given the go-ahead for the crew's presence despite it being "a particularly intense day." Mating had to be precise. Any slippage, sticking, or skewing and a combination of springs and explosive bolts could release the impactor in an unexpected direction. If that happened, the smaller spacecraft would have little time and limited fuel to push itself back toward the comet.

The crane's hook, when fully reeled in, was about 20 feet above the floor—far more clearance than needed for anything Ball had ever built. But the combination of the roughly three-foot impactor, its coffee-table-like stand, the 11-foot-tall flyby spacecraft, and a custom-made lifting fixture made for a close shave.

Two dozen Ball engineers and managers observed from behind retractable barriers of the sort found at airport security. Just five people were allowed in the inner circle near the spacecraft—Galamback, three technicians, and Lorna Hess-Frey, the lead mechanical engineer. Galamback looked like a surgeon with a patient open on the table. "Joe gets this intense look on his face," a colleague said. "He's like my hound dog when he sees a rabbit."

Galamback told his team and the others in the room: "If anybody sees anything that we're missing, say stop. Anybody in this room can say stop."

And then: "Let's do it."

The sound of an electric motor working far overhead filled the high bay. Baldwin, manning the crane control, inched the spacecraft up. Then the crane stopped. The clean-room fans seemed to grow louder. The flyby spacecraft was about two inches too low. Behind the mask, invisible to Discovery Channel cameras, Galamback's mouth hung open. The engineers said it should clear. It wasn't clearing. As Galamback's team huddled, Henderson stood with the Discovery Channel crew, improvising, "We're verifying right now all of the clearances are pristine."

Galamback and the others in the huddle decided to lower the spacecraft a bit, then crank the crane up in a high gear Galamback and Baldwin called "ramming speed."

The crane's electric motor hummed. The flyby spacecraft cleared the impactor. The "high-tech space crane," as Discovery Channel viewers would know the standard industrial tool, had two upper limits. The one for ramming speed was higher.

Galamback's team lowered the flyby spacecraft a few millimeters at a time, checking the interfaces, the fit of thermal blankets, the positions of the explosive bolts. An hour later the two spacecraft were one.

There were no high-fives. Engineers and technicians beamed in silence. There was Deep Impact.

"It was the first time that I actually thought we might be able to do this. That it might actually work," Galamback said.

Someone from the film crew interrupted.

"Are you guys going to at least clap?" the man asked.

On cue, they celebrated.

NOW THAT THEY WERE AN ITEM, the combined spacecraft had to be balanced. The originally proposed Deep Impact had gone through a complete redesign in large part due to tipsiness, which can prove fatal coming off a spinning third stage of a rocket. Now engineers would see how centered the updated design was.

The impactor had been balanced already, solo testing having been necessary because it would fly alone. Smaller than the first OSO and symmetrical, a turntable-like spin balancer worked fine for that. The flyby was another story.

The solar panels and the instruments created air resistance, and turbulence would confound results from a spin-balancer. Early on, the idea was to truck the mated spacecraft 40 miles south to Lockheed Martin Space Systems and have Lockheed spin-balance it in a vacuum chamber. But it would mean halting other testing for a solid week—you can't "utilize the hardware on the floor," as Nick Taylor put it, when the hardware has left the building—plus all the associated marching-army costs. Then there were the risks involved with preparing the spacecraft for shipping and sending it on its way.

Tim Torphy, who had led test planning for Deep Impact, had the idea of using a piece of equipment gathering dust since the 1980s. It had balanced the Earth Radiation Budget Satellite, which had looked like an obese butterfly. Rather than spinning a satellite, it oscillated the satellite back and forth. Using this Ball Dynamic Balancing Machine, Mark III, could save maybe $200,000 and, perhaps more importantly, afford the Deep Impact team another week to test and fix problems in Boulder. Torphy tracked down the machine in a warehouse and had it moved to the Fisher Complex testing area. Then he asked Michelle Goldman to figure the thing out.

Michelle Goldman during structural testing of the Deep Impact flyby spacecraft

Goldman, a Denver native with a master's degree in aerospace engineering from the University of Colorado, had recently returned from an 11-year stint in Tel Aviv, where she calculated the trajectories of Shavit rockets for Israel Aircraft Industries. She arrived at Ball in 2001 and started on the Deep Impact test team, focusing on the design and testing of electrical cables. She rose to lead the testing of structures and mechanisms. Her work encompassed moving parts including the solar panel latches and struts, as well as explosive bolts and springs connecting the spacecraft to the rocket and the two spacecraft to each other.

Goldman beheld the Ball Dynamic Balancing Machine, Mark III. Its boxy footprint would consume half a parking space. It exuded robustness—there were heavy supporting tubes welded at angles and a thick cylinder rising to support a platform the height of Goldman's chin. The whole thing was painted sky blue.

The control terminal looked about as heavy as the spin balancer itself. A key electronic component, a Princeton Applied Research Model 4203 signal averager, was missing, probably sold off years back.

Goldman, an engineer facing a murky problem, set about the task. Were there engineering drawings? Yes, but skeletal. Could they replace the antiquated signal averager? Yes: she found one someone was happy to part with—for $5,000. But however many hours she spent fiddling with the machine and the signal averager, the system defied her. Her bosses decided to call the guy who had designed it.

The guy was Dick Woolley, the man who for 42 years at Ball Aerospace had harnessed his mastery of sophomore physics into engineering bordering on sorcery. Establishing parentage had been the one easy thing about the Mark III. It was still known as the "Woolley Wobbler."

Woolley had retired in 1991. He and his wife Jeanne lived in a quiet con-dominium in Denver, with photos of grandchildren on the walls, Christian reading materials on the shelves, and mountain views out the ninth-floor win-dows. The phone rang. The caller asked if Woolley might be willing to get the Wobbler working.

That the Wobbler would again see the light of day surprised Woolley. That a $300-million mission depended on it shocked him. That Ball engineers couldn't get the machine to work made perfect sense.

Woolley had designed it in a hurry, mostly in his head, 18 years earlier. Nothing remotely like it had existed. But with Woolley behind it, Ball man-agement all the way up to Gabe Gablehouse in the president's office had no doubt the new machine would work.

But it hadn't worked. Woolley had made what he called a "stupid mistake." The bearings were too soft, rendering the Mark III useless for much more than confirming the integrity of the floor beneath it. He scrambled, ordering parts and kludging in an ingenious fix at Cape Canaveral just before the 1984 shuttle launch.

"It was not elegant engineering," Woolley later admitted. "I could design a much better one today if I had to." But in light of the unexpected work it landed him in retirement, he joked, "It's an example of how bad workmanship makes for job security. The rule is never to do anything very well."

In the winter of 2002, the 77-year-old Woolley had his urine drug-tested so he could do a brief, lucrative consulting gig at his old employer.

"I had visions of a kind-faced grandfather-like guy who would benevo-lently come back into the 'good old company' and give the new guys a hand with some of his old equipment," Henderson said. "He hardballed us."

So here was Goldman, an engineer in her late thirties, debugging technol-ogy half her age with a man twice her age. Goldman listened as Woolley, using a mallet as a prop, explained the essence of dynamic spacecraft balance. She found this "cute." She well understood that when Deep Impact spun up, its axis could wobble up to 2.25 degrees. Any more and the mission was lost. Woolley proved to Goldman over a couple of days that, touchy as his Wobbler was, it could be made to behave. Then he went back to retirement for good.

The Wobbler waited another year and a half until, one day in April 2004, Goldman and a small team sat the spacecraft on Woolley's contraption and wobbled it. The mated flyby and impactor proved to be well-centered, though not perfectly. They attached counterbalancing trial weights to even things out—perhaps a pound at first, then down to a fraction of an ounce with successive runs—until the Wobbler had balanced Deep Impact. They replaced the temporary steel balance weights with copper a Chilean mining company had donated with the hope that it would form the impactor's cratering mass (it was, alas, not pure enough). It was time for Deep Impact to run the environmental-testing gauntlet.

Chapter 20

Hell Weeks

Environmental testing is to a spacecraft what Hell Week is to a Navy SEAL trainee. SEAL trainees lug 300-pound rafts, swim against frigid tides, and crawl through mud while instructors invite them to ring the *I surrender* bell, have coffee and a donut and go home. Spacecraft are bombarded by electromagnetic radiation, blasted with noise, frozen and cooked for months, and there is no surrender bell. Environmental testing exposes a spacecraft's weakest links in order to fix them and make them stronger.

Six people rolled the mated Deep Impact spacecraft out of the clean room, through the changing room, out the back door, uphill across a parking lot, and into the test area in another part of the Fisher Complex. Deep Impact was placed in an anechoic, or echoless, chamber. Behind multi-story steel doors, the blue-black room featured two-foot foam spikes stabbing from its walls and ceiling. Inside, the spacecraft would endure electromagnetic interference and compatibility testing, or EMI/EMC, as the Deep Impact team called it. The test blasted the spacecraft with radio waves, and the spacecraft blasted radio waves back.

Preparing for
electromagnetic testing

Jon Mah, the young test engineer, volunteered to plan and oversee the test. He learned what his boss Nick Taylor described as a "black art." Mah had to make sure the spacecraft's radio signals didn't disturb the rocket and vice versa. Self-compatibility testing ensured Deep Impact's own systems made good electromagnetic neighbors.

Mah was distinguishing himself as a budding talent. He learned quickly. When he had questions, he was polite but insistent. Test engineers like Mah relied on the specialists who had built the computers, the attitude control system, the telecom system, and the instruments. Mah sent e-mails, and if the specialists didn't answer quickly, he tracked them down at their desks "You don't wait around for somebody to reply to an e-mail or something. If you've got to figure something out, you find that person," Mah said. "You've got to work. What're you going to do, like, sit there and think about it?"

Mah learned that testing for self-compatibility required him to turn on and off all sorts of electronics on the spacecraft, not just the radios, because whereever electricity flowed, electromagnetic waves emanated. He learned that bombarding Deep Impact's antennas with their preferred frequency would fry communications electronics listening for a far fainter signal, so he had to block those frequencies out of the test. Then one night, with the spacecraft in the foam-spiked cavern, Mah took control of Deep Impact as specialist engineers ran the test chamber's electronics.

"When I look at reviews, it says an associate engineer can do limited testing—essentially needs a senior engineer to hold his hand," Taylor said. "And I'm thinking, he's in there on his own, running both spacecraft at the same time."

Providing the independence to let talent flourish had been a part of Ball Aerospace's culture since the beginning. Bill Frank, one of the company's early engineers, explained it: "You gave him a job and left him alone. And said, you know, this is your responsibility. You've got to get it done. And if you need help, you've got to come and get that help. But we didn't try to tell them how to do things or check on them all the time. I think Ball's willingness to give a guy more responsibility than he'd ever had before was a major factor in acquiring engineers and keeping them."

THE SCREAM OF A ROCKET in flight can damage its payload. For deep-space missions, NASA demanded acoustic testing. Ball Aerospace had never needed a special chamber for acoustic testing. The initial plan was to pay Lockheed Martin to test Deep Impact in their acoustic chamber after Lockheed checked the spacecraft's balance. The Woolley Wobbler's resurrection had taken care of balance, though. Perhaps there was another way to do acoustic, too. Someone suggested calling Maryland Sound.

Maryland Sound International, based in Baltimore, set up audio systems for presidential inaugurations, Times Square New Year's Eve celebrations, papal visits, and big concerts. They had designed and installed the speakers for a Pink Floyd show on a barge in the middle of Venice's Grand Canal. Tucked at the bottom of the company's Web site was a link to something called "Top Secret." It read: "MSI is cleared to conduct highly classified acoustical testing assignments on behalf of both public and private organizations. Obviously, we can't talk about them...."

Bob Goldstein had founded the company in 1966 to do the sound for a Frankie Valli & the Four Seasons tour. Among the Top Secret jobs Goldstein could talk about included his firm's acoustic testing of massive DirecTV satellites, cruise missiles, and the Mars Exploration Rovers. This was odd fare for a high-end audio house, Goldstein admitted.

"We've specialized in weird our entire existence," Goldstein said. "Our motto is, if it's impossible, we'll do it. We're known as the Doctors of Strange."

Loudness roughly doubles with every three decibels. Normal conversation is about 60 dB. Permanent hearing damage is a real risk with long-term exposure to 85 decibels, the volume of lawnmowers and crying infants. Rock concerts sustain about 105 dB, with potent guitar riffs reaching 120 dB. Even brief exposure to that damages ears. Nature knows this. People experience pain when sound exceeds 120 decibels. Deep Impact's acoustic testing would happen at 143 dB, thirteen times louder than a rock concert. Such volume created a breeze.

"It's everything, all at once, flat-out, for a sustained period of time," Goldstein said. "Most rock-and-roll sound systems would burst into flames," a fact he said he learned from experience. Goldstein once donned triple hearing protection, stood in the vortex of speakers arranged for a satellite acoustic test, and asked his guys to crank it up to about 135 decibels. He felt like his bones were being jarred to pieces.

The Deep Impact acoustic test was planned for mid-May 2004. Ball and Maryland Sound negotiated the date. In June, the speakers were booked for the Daryl Hall & John Oates Rock N Soul Revue Tour.

"From my perspective the interesting thing was, although the schedules say, 'We're going to do the test on that day,' it never happens. There's a delay here or there's a delay there," Ball test manager Nick Taylor explained. "But we

Acoustic testing

were running into concert season. So these guys have got contracts out with Rod Stewart or Hall & Oates or whoever. Now we say, 'We need to do this test,' but they have a window of opportunity, and if we miss that window of opportunity, Rod Stewart or Hall & Oates have priority over Deep Impact."

During three days beginning May 10, Maryland Sound technicians set up a 10-foot-high ring of speakers and tested their audio equipment near the vibration tables. They had long hair and wore Black Sabbath and Harley Davidson T-shirts. Genuine roadies had set up shop in a place of higher engineering. At one point, the sound of a solemn bell reverberated from the test area, followed by a heavy guitar riff. The crew was testing the audio system with AC/DC's "Hell's Bells."

Ball technicians and engineers rolled the mated Deep Impact spacecraft into the speakers' vortex. The Maryland Sound techs cranked up the volume on a track that sounded like a vacuum cleaner, if Pratt & Whitney had made vacuum cleaners.

The spacecraft was awake and operating at the time. Taylor waited out the test in a nearby corridor behind two sets of double doors, "and it still sounded like a freight train coming right at you." Jim Baer, the Deep Impact optical engineer, was 50 yards away and behind steel doors. "You couldn't hear yourself think, and that's with the headphones," he recalled. The spacecraft, though, came through it just fine.

DEEP IMPACT PERCHED on a vibration table, a bastardized subwoofer packing double the punch of the one that had cracked the big telescope's mirror a year earlier. The spacecraft, removed from the clean room, had been bagged in a

The Deep Impact spacecraft, enveloped in antistatic wrap, is lowered into the thermal-vacuum chamber known as Brutus.

layer of static-resistant silvery plastic and shrouded with a second bag dangling from a crane hook. The table, oscillating up to 2,000 times a second, the highest C on a piano keyboard, delivered enough force to shake parts off a new car. The spacecraft survived.

Now it faced Brutus.

Brutus, a massive thermal vacuum chamber, looked like a propane tank for Paul Bunyan's gas grill. Brutus could bake its contents to 320 degrees Fahrenheit or freeze them down to minus-320 degrees. It was an oven; it was an anti-oven. Deep Impact would experience only moderate temperatures by Brutus's standards: minus-290 and just over 100 degrees on the warm end.

Ensconced in its silvery protective wrap, Deep Impact was lowered by crane into Brutus's 20-foot maw. Technician Alec Baldwin climbed a ladder to ratchet tight lines supporting the solar panels. Designed for zero gravity, the combination of Brutus's extreme temperatures and terrestrial pull could warp the panels' titanium mounts. Baldwin and colleagues plugged in cords through which power and the model universe would stream.

The crane replaced Brutus's 15,000-pound lid, and the chamber's plumbing churned for a week, squeaking and moaning as it pushed fluids and gases through a network of thick white pipes leading, among other places, to a three-story liquid nitrogen tank outside the building. Brutus chilled and drained itself of air molecules until its innards were about as cold and empty as anything on Earth can be.

The spacecraft spent 28 days in the bowels of Brutus during June and July. The team testing the spacecraft worked three shifts, around the clock, made easier only because they knew thermal vacuum testing, when successful, was one of the last steps before a spacecraft shipped off to Cape Canaveral.

John Valdez, responsible for making sure Deep Impact maintained its cool, was most under the gun. He watched temperature data coming from the spacecraft to see if it jibed with his mathematical models. Engineers from the subsystems spent hours at the test consoles outside the chamber, inspecting their own numbers as Deep Impact, again convinced it was flying through space, revealed its every thought via the umbilical cabling.

Henderson, in addition to his management duties, took on the role of program caterer. He kept a supply cabinet and refrigerator stocked with Costco snacks—sodas, chips, fruit, candy. He brought in food from Wahoo's Fish Taco, KT's Real Good BBQ, Red Robin, and Abo's Pizza. Food had a mystical ability to refocus an engineer's mind from, "Man, I've been working for 17 hours and this just sucks," to "Snacks!" Engineers tired of take-out could be found in a driving rain at three in the morning, flipping pork chops or burgers on an outside grill near the towering liquid nitrogen tank.

One shift bled into another as delayed tests pushed back the starts of ones that followed. It meant those arriving at 3 p.m. for the second shift might wait hours before they could start their work, their 11 p.m. quitting time pushed back to two or three in the morning. People tended to work the better part of two shifts, six or seven days a week as Brutus warmed, froze, thawed, froze and thawed the spacecraft again. Engineers became so familiar with the spacecraft they could tell what it was doing based on the current draw alone.

The JPL and Ball engineers building Deep Impact were finally uniting into the high-performing team Brian Muirhead and others had envisioned. This had happened gradually, organically, shaped by the same context driving all high-performing teams: strong individuals, mutual respect, trust, and, most importantly, a monumental challenge plunging the shared enterprise into adversity.

At the very top, cultural differences and financial pressures never eased, and relationships among those on the business end of Ball and JPL remained tense. But in the build-it, test-it, ship-it technological trenches, a *Band of Brothers* mentality had taken hold. Hitting the comet and succeeding in space were important to some; solving tough problems motivated others. But everyone wanted the enterprise to succeed, and no one wanted to disappoint the team.

As one described it: "I wanted to hit the comet, I wanted Ball to be successful, I wanted to show JPL we could do this. But you know my biggest motivator—I didn't want to let anybody down on my team."[1] Said another: "The fact was that we all got along so well and we all supported each other.

You didn't want to be the guy who threw his hands up. Because you were going to screw the rest of the team. I really think that helped get us through it—the fact that everybody really not necessarily cared about the project, but about not letting the team down."[2]

Mah had similar sentiments: "It goes back to where it's like, if I don't do it, I'm gonna screw my buddy, and we're all buddies here."

One evening, Mah reviewed numbers streaming from the spacecraft trapped inside Brutus. He had been doing this for many long days straight. His boss Taylor walked up. The veteran of several such test marathons sat down next to his young colleague. Atop a concrete floor, inside cinderblock walls and steel doors, under overhead cranes, beside the giant Thermos grunting and hissing away, Taylor turned to Mah and added a human voice.

"You're going look back on this, Jon, and you're going to say, 'Those were the good old days.'"

Mah, who had seen little but Brutus and his bed in two weeks, weighed the idea for a moment and turned from his screen.

"Fuck you, Nick," he replied, with only the faintest hint of a smile.

Chapter 21

Tag Team and Tobacco

By the time the spacecraft underwent its hell weeks, John Marriott, although still Ball's project manager in name, had been promoted to over-see a handful of missions in addition to Deep Impact. Henderson retained the title of deputy project manager, but the scope of his role expanded to cover Marriott's duties.

A year earlier, in early 2003, Henderson had prepared a list of possible cost cuts and submitted it to the project's JPL managers. At the bottom, as an afterthought, he had jotted down "overtime pay." Aerospace contractors paid their salaried engineers overtime. At Ball, overtime kicked in at 45 hours, and was intended to make up for months-long stretches of weekend work and the sacrifice of personal lives. There had been unpaid overtime from the beginning of the mission, but not a lot—mostly lone engineers working late because of deadlines or personal commitment. JPL, the Cal Tech/NASA nonprofit, by policy paid no overtime. JPL's managers had seized on Henderson's offhand suggestion and eliminated overtime pay.

The decision had surprised Henderson, and he had felt terrible about it. Ball engineers were livid. They viewed the cut as "an extremely poor decision made by people who had never had to work those kind of hours themselves," as one put it. Henderson had leveled with the team. "He just said, 'Whoops, I made a mistake. Sorry everybody.' And it was thousands of hours of people's lives," as another remembered it.

The move saved millions of dollars on an effort involving 20 to 50 percent more workers than had been planned. It may well have been the difference between cancellation and survival.

NASA Headquarters continued to stalk the mission. Weekly status-review teleconferences, for which Henderson assembled and e-mailed off 100-page PowerPoint documents, delivered constant doses of angst. Colleen Hartman, who managed NASA's Solar System Exploration Division for Ed Weiler, read aloud the Discovery Program cancellation provisions at the start of one virtual meeting. On Mondays Henderson spent an hour apprising Ball Aerospace boss David Taylor of Deep Impact's status. "Deep Impact has got to work," Taylor told him, over and over. "It is absolutely critical for this company. You have to make this work."

Among Henderson's roles was to act as an umbrella over engineers, shielding them as best he could from such up-and-out storms so they could focus on getting the spacecraft built and tested. In Henderson's world of budget and schedule, there were threats, yelling and fist-thumping. NASA space science chief Ed Weiler and his deputy Hartman, both PhD astrophysicists, shouted at the esteemed astronomer Mike A'Hearn and, in a twist on a parent-teacher conference, even visited the president of the University of Maryland to explain the state of affairs. A'Hearn, with time, noticed a certain cadence to Weiler's hollering, and took it to be, at least to some degree, role play. Hartman's yelling he found more terrifying.

Lindley Johnson, a retired Air Force Colonel, came on as NASA's day-to-day Deep Impact program executive in late 2003. His job was to make sure Deep Impact "was staying in the box, as we say," but also to advocate for the mission at NASA headquarters. He walked a fine line between "bad cop" and going native.

To him, Deep Impact seemed forever on the brink of cancellation. The mission's leaders came in roughly every six months to plead for more money. By early 2004, Johnson had a draft termination letter for Deep Impact waiting on his computer hard drive. He might have sent it were it not for a few factors.

One was the Contour disaster, which had already erased a NASA Discovery Program comet mission. Second was Weiler's and Hartman's departures in mid-2004. Weiler's belief in the mission had seen Deep Impact through its darkest days, but his patience had limits. Deep Impact's fourth or fifth request for millions more could have done it in, Johnson felt.

"Missions are always back at least once, twice, three times," Johnson said. "But coming back five times—that is pretty gutsy."

Fortunately for the mission, Deep Impact had yet to become a personal annoyance to Alfonso Diaz, Weiler's successor, or Orlando Figueroa, Hartman's. In addition, Johnson said, the Discovery Program's rejection of all proposals in 2004 freed millions of dollars to shore up Deep Impact and Mercury Messenger. And finally, Johnson and his bosses figured they'd see the comet mission again if they cancelled it now.

"The science was so compelling that we fully expected it would get reproposed," he said. "That would just end up costing much more and delaying it however much longer."

Still, the pressure on the ground was relentless. Henderson, a first-time project manager suddenly atop Ball Aerospace's biggest program, often bolted upright in bed at three in the morning, thinking about one problem or another. He lived in a haze, immersed in the sleep deprivation that had long beset his engineers. He resorted every few nights to Tylenol PM, which afforded him seven hours of sleep. He was at Ball every day. Often on Saturdays Henderson brought his three-year-old son Reed along, snapping him into the smallest bunny-suit smock he could find and tying its arms around the child's back as a sort of impromptu straitjacket for wear during an 8:30 a.m. "tag up" meeting in the clean room. Henderson led the meeting as Reed sat quietly, not far from the spacecraft, patiently waiting for the reward of whiteboard markers and drawer toys upstairs in his dad's office.

Contributing to Henderson's stress was another management change at JPL. In late 2003, John McNamee was promoted to oversee several JPL planetary projects. Rick Grammier replaced him as Deep Impact project manager.

Grammier was the son of a textile executive and spent his formative years in Roswell, New Mexico. Before he was born, Grammier's mother had worked with doomed monkeys sent aloft on V-2 rockets to pave the way for the Mercury astronauts. The books of his youth included science fiction such as Robert Heinlein's *Glory Road*. Like McNamee, Grammier had come to the space business indirectly, arriving at JPL in 1988, the same year as McNamee. McNamee had leveraged business experience into success in space; Grammier did the same with his years in the U.S. Army.

Following his graduation from West Point in 1977, Grammier had been a U.S. Army captain leading what he would, decades later, only refer to as "special assignments." He retained the quiet confidence of a man who learned to assess a situation and make life-and-death decisions quickly. "You learn that people are your most important asset and you've got to recognize the strengths and weaknesses of those people, and how best to use those to your advantage and the team's advantage," he said.

Now he was a 48-year-old space manager who publicly described Deep Impact's challenge as "hitting a bullet with a bullet while watching the collision from a third bullet." He had been deputy director of solar system exploration at JPL when lab director Charles Elachi told Grammier he would be taking over Deep Impact. Grammier had been on a review board keeping close tabs on the mission. He knew what he was getting into. Deep Impact was, he said, "the biggest challenge I ever had in front of me."

Grammier called an all-hands meeting. He talked for a while about the work ahead. Then he said, "And you know, we really have a great mission here, and the most important thing is for everybody to have fun."

Carl Buck, the JPL manager keeping tabs on the Ball effort in Boulder, recalled the comment itself being "the last fun or funny thing that happened on the project."

Grammier was, as Lindley Johnson put it, "a hard-ass."

"I mean in a good way," Johnson elaborated. "He wasn't abusive or anything like that, but he made people clearly understand what they were signed up to deliver and made them abide by it."

Just as there had been two John Marriotts, there were two Rick Grammiers. One was a charming, soft-spoken father of four boys and a girl. The other was a commander on a mission. From the beginning, he played no favorites, treating the Ball and JPL teams with equal toughness. He spoke softly and carried a club. He was so organized that Buck received e-mails weeks ahead of review meetings saying, "Carl, here's what I need you to provide." He knew precisely what NASA reviewers were looking for, and had managers set about "gilding the lily and polishing the cannonball," as Henderson put it, in areas Grammier knew scrutiny would be toughest. Rather than bury bad news, he preempted criticism by opening such meetings with frank discussions of problems and things the mission had yet to take care of.

Grammier brought Keyur Patel in as his deputy. Patel, 41, had emigrated from Bombay as a boy and been at JPL since a 1985 summer internship during his undergraduate aerospace engineering studies at Cal Poly Pomona. His intensity overshadowed even Grammier's. "I take it personally," as Patel explained it. "You put me on a job to get it done. I'm going to get it done and make sure it works. You have to personalize it at some level. If you artificially remove yourself from the success of the project, you're doomed to fail."

Patel spoke less softly and carried a club. His management style bordered on hostile, and he was brutally honest, like Marriott. He pushed his JPL team and Ball hard. Mike Hughes, the JPL attitude control engineer, described life on Patel's team as "like working for a military commander that you trust."

"In the fall of 2003, we just knew: there's no way out of this. We didn't have the top cover," Hughes said. "We didn't have the management leadership to help us out of the situation we were in. I went in there and sat with Keyur and said, 'Here's what we've got to do.' He's like, you got it. We're going to call this guy up. We're going to do this. And if they didn't get together, Keyur would figuratively grab them by the scruff of the neck and pull them together. He had an amazing ability to focus in on the right problem to support his engineers and use his management chutzpah, his moxie, to pull in whoever he needed to into that room to solve that problem."

Ball managers respected Patel as a bright, action-oriented counterpart. "Keyur was the kind of guy who said, 'We need to make a decision,'" one explained. "Everyone around him said, 'Oh, we don't have enough information to make a decision.' He said, 'We need to make a decision. We might make the wrong decision. If we make the wrong decision, we're going to do something to recover. But we need to make a decision.'"[1]

Others at Ball chafed at what Henderson diplomatically described as Patel's "aggressive management style."

"Keyur's a real butthead," Marriott said. "He's just death and destruction. This guy could care less about people. They're a commodity. When he burns 'em out, he just gets new ones. No respect at all for another person. Ruthless."

McNamee had been in Boulder more than Muirhead; neither had been there enough, in Grammier's mind. Grammier and Patel alternated weeks in the third-floor Fisher Complex cube farm. Henderson faced a tag team of two of JPL's brightest and toughest managers.

In Henderson's office, Patel would holler to the point that Marilyn Morris, the administrative assistant, feared for her boss's safety. "Do you want me to come in? I can maybe break it up," she asked Henderson once Patel moved on. When in Pasadena, Patel called Henderson's cell phone every evening, sometimes at 5 p.m., sometimes at 10 p.m., with a list of things he wanted done.

Once, during the spacecraft's weeks in Brutus, Patel barked at Henderson in the Ball manager's office, this time about the accuracy of John Valdez's thermal model for the spacecraft.

Henderson took a deep breath. He explained that the model would do the job. Either way, in the middle of thermal-vacuum testing, with Valdez and his helpers working around the clock, there was no way to rework the numbers now, he said.

Patel fulminated, cussed a blue streak, accused Ball of cutting corners and delivering something short of what it took to succeed on a deep space mission. Then he stopped and leaned back into his chair. Henderson braced for another barrage.

"So," Patel asked. "Where do you want to go for lunch?"

PATEL AND GRAMMIER, LIKE HENDERSON, were under enormous pressure. Both were certain that, if Deep Impact failed, JPL would bear the brunt of the embarrassment, as it had with the Mars failures despite Lockheed Martin's complicity. Both men knew many at JPL and NASA Headquarters harbored serious doubts the mission could be pulled off. Both had committed to spending half their lives away from families—Patel had daughters aged one and four—to get Deep Impact to its date with the comet. As they came to understand the true state of affairs, they tempered their expectations to simply getting the spacecraft to the launch pad. The rest they could somehow figure out on the way to Tempel 1.

The spacecraft was physically solid by mid-2004, six months before launch. Frank Locatell, a former JPL senior engineer turned consultant, dropped at Ball for hardware walk-downs. The man was paid to nitpick. He found far fewer nits on Deep Impact than on any spacecraft he had ever seen. He put his arm around Lorna Hess-Frey, the Ball engineer largely responsible for the design, and told her, "This is beautiful hardware."[2]

Brutus exposed only one real weakness—the impactor's propulsion tubes were prone to freezing. Valdez tweaked the heaters, and Alec Baldwin re-wrapped the stainless steel tubes in thermal blankets. They put the impactor back into Brutus and it checked out fine.

Deep Impact, though physically tough, was still unfit for its mission. To hit the comet, the spacecraft had to read from a hierarchy of scripts. A command to turn and follow the comet triggered scripts embedded in attitude control, instrument, thermal, power and communications systems. Each had been independently tested. Rather than a single actor, Deep Impact was a flying theater troupe having to learn all the roles and then perform them in flawless ensemble.

Yet Deep Impact was incapable of learning. It could only act and explain itself. So it was up to the spacecraft's creators to diagnose, reprogram, test, up-load changes, and run the scripts again and again.

They tested on the four test benches and also with the spacecraft itself plugged into Sean Esslinger's universe-in-a-box. The test benches deceived Deep Impact through launch, cruise, and the encounter with Tempel 1. It wasn't enough to launch or cruise or hit the comet with everything working perfectly. Deep Impact had to prove it could shoot straight with a hand tied behind its back. Testers devised simulated disasters—power system failures, temperature spikes, computer memory corruption, antenna quirks. Systems supposedly redundant had to prove it, so they ran tests with one or the other communications amplifier out of commission, a gyroscope on the fritz, a cold thruster, a blinded star tracker, a comatose computer. They tested how Deep Impact would react to a solar storm, a banana-shaped comet, even a dead im-pactor. And they tested combinations of the above.

Such tests went by various names: mission-scenario tests, mission-readi-ness tests, operational-readiness tests, performance tests. Whatever you called them, they weren't going well.

With all the threats of cancellation and financial struggles, the mission leaders had had a choice. They could either pay for the spacecraft or pay for mission testing. Working against testing was the benefit of enhancing the hardware's pure physicality, which would remind NASA managers of what its dollars had bought much more vividly than untold pages of mission-test documents. Henderson's first act as Marriott's deputy had been to cut the mission-test team from six engineers to two. The project needed the money elsewhere. If the mission got cancelled before Deep Impact was built, there'd be nothing to test anyway.

A few voices dissented. If you build a spacecraft but don't test it properly and it fails, why build it in the first place? Bill Frazier made this point, and lost his job over it.

Frazier, the chief systems engineer who had helped spot some of the initial spacecraft design's most glaring weaknesses, considered a lack of mission testing among Deep Impact's gravest risks. The proposal team had suggested that just two people would work on writing test scripts and running them through test-bench computers for a couple of months to make sure the spacecraft could do what Ball and JPL said it would. Frazier figured it would take 10 engineers working for three years to get it done. Marriott understood, but there was simply no money to pay for the extra bodies. So Deep Impact procrastinated in the worst way, cramming for its toughest tests in the last months of the mission. Frazier, adamant, had been shown the door in early 2003, evicted from the mission he had done much to shape.

The true price of the one-year launch delay was also becoming clear. The extra time had saved the mission. But, at least as Grammier saw things, the delay now threatened to kill it. The idea all along had been to sharpen mission operations during Deep Impact's leisurely 18-month cruise around the sun. Now there would be just six months in space, and as far as Grammier was concerned, mission operations had little to show for the extra year on the ground. The spacecraft was a superb physical specimen, perhaps, but unpredictable and, ultimately, unreliable. When Patel and Grammier joined the project, Deep Impact was habitually missing the comet in tests, and time was short.

Surprises also consumed managers' attention. Grammier noted that the spacecraft lacked lightning protection. If lightning struck the nose of the rocket while Deep Impact waited on the launch pad, the bolt could fry the spacecraft. Ball had never bothered with lightning protection. JPL insisted on it. Grammier launched a crash program to make the wiring changes.

Then there was tobacco. The idea of ceremonial tobacco riding along on the impactor spacecraft began with a casual conversation between a NASA public-outreach person and a man named Ron His Horse Is Thunder, who was president of Sitting Bull College in North Dakota and the great-great-great grandson of Sitting Bull himself. NASA had learned that some Native American groups disliked the idea of blasting a comet; Ron His Horse Is Thunder explained why.

"If you take a sample, you should leave something behind, and not just the spacecraft," he advised the NASA person. "If you could leave a piece of tobacco, that would be wonderful. Other Indians would hear about it and know you had spoken to an Indian. It would show an interrelatedness. A connection."[3]

Deep Impact would take no samples. It would carve out a sizable scoop of comet, take pictures, and gain knowledge, though.

NASA public-outreach director Shari Apslund was enthusiastic about the notion of a tobacco offering. She envisioned events in North Dakota "culminating with a Spiritual Advisor performing a ceremony," meaty press coverage, and "a lot of good feelings about the mission among Native Americans." A Deep Impact public-outreach staffer agreed it should be considered. E-mails from both landed in Henderson's and Grammier's inboxes.

It was mid-2004. The project had taken zero days off since 2003; many engineers and testers worked six and seven days a week. A seven-day work-week no longer counted as a "week" in the traditional sense. The intervening weekend gone, it became, at minimum, a 12-day workweek. The spacecraft was going through environmental testing. Mission testers were rooting out quirks. NASA Headquarters was riding herd on the mission. No one was sleeping enough. A bit of tobacco flared into a big deal. Would they have to re-balance the spacecraft? Or write reports explaining why they didn't?

Henderson assumed the directive had come from NASA Headquarters. Indeed, Native Americans had complained to the space agency about the cultural tone-deafness of missions like Deep Impact and Stardust (the comet dust sample-return mission that had successfully flown past Wild 2 in January 2004) for violating a universal law of reciprocation. Henderson responded that, as long as well-sealed tobacco could be mounted before the impactor's solo time in Brutus in July, it shouldn't be a problem. The "Send Your Name to a Comet" CD had been similarly encased and mounted, so there was precedent. He did suggest, though, avoiding the press, at least until after launch. "I don't want this to trigger a flood of other religious, scientific and educational institution requests to carry their artifact to the comet, too," he said.[4]

Grammier concurred, but asked: "Is this really needed/necessary? i.e., If we don't do it are there really any repercussions?"[5]

Both managers put the tobacco out of their minds. Unbeknownst to Grammier and Henderson, Mike A'Hearn's own public-outreach people had neglected to mention the tobacco idea to him. No longer a space-engineering neophyte, he responded forcefully when it came to his attention.

"As you know, we are in a continuing battle to control both cost and schedule, and, although some changes may seem trivially simple, any perturbation can have significant ripple-through effects," he wrote in a letter to the Deep Impact education and public-outreach director, whose University of Maryland office was a few feet away from his.

Changes meant repeating environmental tests, he continued, and at minimum there would be paperwork. "I do not intend to give permission for any changes absent extraordinary arguments," he finished.[6]

No SUCH ARGUMENTS CAME. Testing forged on. Problems cropped up. They were duly noted in hundreds of TAWS, PAWS, MDRs and other document-vehicles designed to transport descriptions of Deep Impact's problems to those who would address them and the managers keeping track.

On the first day of October, Ball opened the clean room to the media, to afford them a look at the spacecraft before it was sent off to Florida. The local press, mainly, showed up. Reporters were treated to a look at Deep Impact as well as to mind-blowing analogies ("It's like hitting a charcoal briquette with a BB from six miles away, against a black background, at night," one engineer explained). To the untrained eye, the comet hunter was a curious monolith wired to a bank of computers of mysterious utility. The spacecraft was in the middle of a 30-hour test, the visitors were told. How the tests were going, no one thought to ask.

Deep Impact looked fully formed. The message of "we're shipping" was understood by the reporters present to mean "we're done." Deep Impact, though, was shipping with the equivalent of "some assembly required."

The stories that emerged focused not on what Deep Impact actually *was*, but rather what Deep Impact was to accomplish—the beginning of what became a mountain of breathless press about the mission. "NASA hopes to hit a cosmic fastball moving through our solar system next year with a high-tech bat made in Boulder," the *Denver Post* story the next day began. "Built by Ball Aerospace in Boulder, the Deep Impact mission will line-drive an 800-pound copper bullet deep into a comet somewhere in the solar system's infield."[7]

It was one of Ball Aerospace's few moments in the spotlight. The public had little idea that most NASA spacecraft were in fact the work of aerospace

A spacecraft fully formed in body only

The Deep Impact deputy program managers Keyur Patel of JPL, left, and Monte Henderson of Ball Aerospace add the Deep Impact mission sticker to the door of Brutus with the completion of testing in Boulder.

and defense companies such as Ball Aerospace, Lockheed Martin, Boeing, Northrop Grumman, Orbital Sciences and hundreds of subcontractors. At best, contractor names ended up at the bottom of NASA press releases, the patron NASA taking credit for the art.

The spacecraft was unplugged from its computers a few days later. On October 10, three trucks showed up outside the Fisher Complex. Two were soon packed with accoutrements such as the Woolley Wobbler and the computers to test the spacecraft.

The third truck, an unmarked Barrett Moving and Storage air-ride flatbed, carried the spacecraft. Baldwin, Joe Galamback, and others drained the battery, sheathed the solar panels in aluminum covers, and double-bagged and stowed Deep Impact.

At 3:30 a.m. on October 14, 2004, Deep Impact rode off into the darkness. Henderson watched it go. It took four days to drive to Florida, with escorts ahead and behind the precious cargo. Inside, the temperature was 68 degrees and continually pumped with nitrogen to keep humidity down. Shock sensors recorded every bump from Colorado to the Kennedy Space Center.

Part IV

Taking Aim
for
Tempel 1

Chapter 22

Faulty Liftoff

Florida's September rains had given way to sun, with temperatures in the eighties. Deep Impact engineers coming from more than a mile above sea level took up surfcasting and surfing proper; they deep-sea fished and golfed and jogged on the hard-packed sands of Cocoa Beach. They spent time in bars. And they worked on their spacecraft, whose life had changed very little. Deep Impact spent its days—and nights—being deceived into thinking it was flying.

The dual spacecraft was in a clean room at Astrotech, a company whose facility served as a Holiday Inn for visiting space robots. Launch was less than three months away. Deep Impact was not ready. In five comet-encounter tests before shipping, the impactor had met its target just once. The encounter script—command sequences directing every move both spacecraft would make from six hours before they separated through the fateful meeting with Tempel 1—still needed thousands of hours of testing and refinement. But the comet wouldn't wait. Deep Impact had to start its big show in space with the ending unwritten, and learn its most important lines en route.

For JPL, it was nothing new. Deep space missions lasted months and years. To fill the hours, technicians sent software updates across millions of miles, then tested, resent, and retested. Once the spacecraft reached its destination, whether orbiting Saturn or roving on Mars, initial discoveries led to new plans to program into a spacecraft via electromagnetic waves traversing space. But it was still risky. If a full on encounter test revealed a hardware problem, there would be no fixing it in space.

From Ball's perspective, launching an incomplete spacecraft was a foreign concept. "Typically you go to the launch site and it's quiet time. You're just doing minor prep work before you go into the rocket," Henderson said.

A team of 20 Ball and another 10 JPL engineers and technicians trained the spacecraft for its journey. They worked two shifts most weeks, and occasionally around the clock. There were no weekends—the fishing and fun-in-the-sun happened before or after work.

David Wilson, the electrical specialist who had managed the impactor spacecraft team, was back on the power system, responsible for trickle-charging the flyby spacecraft's battery. He tended to work second shift. A couple of drywallers kept showing up near his condo in the morning; they stood in the surf, cast into the waves, and drank a 12-pack each. Wilson walked down and talked to them one morning, and again the next, then bought flycasting gear and joined them, though without the 12-pack. Then he'd filet his fish, eat some of his catch, drive to Astrotech, and tend to his spacecraft.

Jon Mah, the spacecraft tester, worked long days and bar-hopped long nights. "I've been to so many bars in Cocoa Beach, people know me by name," he said. Tom Golden, the software engineer often joining Mah, once responded to a Sunday evening page with the thumb-keyed words: "I'm much too drunk to be able to help you with anything right now."

Others commuted. Tom Bank dropped into Cape Canaveral for multi-week, work-only visits. On the Deep Impact team since 2000, Bank, 47, had been a floater, helping out with instrument performance, doing computer modeling for the attitude control team, troubleshooting star trackers. In 2003, he had become Ball's lead mission systems engineer. Bank's job was to make sure the spacecraft learned its lines and could rehearse them convincingly.

Bank had joined Ball Aerospace at age 30, straight from an undergraduate electrical engineering degree at the University of Colorado, where his dad had worked with wind tunnels as a research assistant. Bank had graduated from Boulder High. He accounted for much of his 20s with one sentence: "I'm one of those guys who traveled a lot and pounded nails." He had planned on spending just enough time at Ball to pay off his student loans and save $10,000 to travel the world.

"My student loans have been long since paid off. And I've long since saved $10,000. And I still haven't gone around the world, and I'm still working at Ball Aerospace," Bank said. "Why? Because it's fun. I'm not willing to turn my back on it. Because every day I go to work and work on cool stuff." On the walls of his garage, next to the workbench and below the camping gear, hung NASA Group Achievement Awards.

Before Bank accepted the mission systems engineer job, he had a conversation with his wife. "This is irrational, and I hate to admit it, but I really want to hit that comet," he began. "And I think we can do it. I'm not going to be able to do it working 40 hours a week. In fact, I can't do it working 60 hours a week. I've got to give it my all. And the only reason I can possibly put up with this is come July 4, it's over. It's done. But I need you to support me."

They had a three-year-old boy they had adopted from Bolivia; she worked part-time as a chemist. But she said OK. And Bank plunged into the void: nights, weekends, coming home and collapsing in bed, his pager beeping at 2 a.m. Months later, at their annual "hat party," Ginny Bank wore a black veil with a sign that read "Deep Impact Widow."

In Florida, at about 3 a.m. in the Astrotech control room, Bank observed as Craig Long, the man described as a "badass testing machine," ran the spacecraft.

As a rule, every instruction flowing through the Deep Impact spacecraft was supposed to have passed muster on one of the four test benches at Ball and JPL. But sometimes a script that *should* work stumbled on a test bench. *Maybe it was a bug on the test bench*, an engineer might think. Given more time, he might undertake a prolonged investigation, reconcile the test computers with the spacecraft, and try it again on the test bench. As pressure mounted, though, the shortcut was irresistible. *Why not just try it on the real deal?*

This is "a big no-no," Henderson explained. "But it's the kind of thing that, when you get closer and closer to launch, you start taking a little bit bigger gambles because your window is going to close on you."

Bank and Long, through windows separating the control room from the spacecraft at Astrotech, watched Deep Impact do what appeared to be nothing. But the spacecraft was a like a termite mound, with extraordinary activity pulsing beneath the quiet shell.

"How are things going?" Bank asked.

"Oh, we're tumbling out of control and I have no idea what's going on," Long answered.

Bank looked at the monitors. He had come to know the spacecraft like an old dog. A quick glance at the telemetry pumping out of the spacecraft told him something was wrong, even if he couldn't place it.

"Well, what are you expecting to happen?"

"I don't know," Long said. "Go ask him."

He gestured to Greg Horvath, a JPL engineer. Bank asked him what might be happening.

"We're tumbling," Horvath said. "We've got a queue of responses lined up. We're not sure what's going to happen next."

"What are you hoping to get out of this?" Bank asked.

"We expect to recover," Horvath said.

"How long will it take?"

"Not sure."

Bank walked back over to Long. "Are we in any danger of damaging the spacecraft?" he asked.

"Naw," Long said.

So Bank watched. Deep Impact remained a stone rolling down an endless hill. It was never going to recover, Bank concluded. He walked over to Long and said, "We're done."

THE ABORTED TEST INVOLVED one of the most contentious, delayed, and, in Ball's view, mission-threatening systems of all: fault protection.

The fault-protection system was software. It was a hand-me-down from Deep Space 1, which had been a dolled-up version of the system in Mars Pathfinder.

JPL had been flying fault-protection software since the Voyager program in the mid-1970s. It took the pulse of a spacecraft's systems, noted anything out of the ordinary, and, if it deemed the problem serious enough, took over the offending subsystem or even the entire spacecraft so the machine didn't hurt itself.

Ball Aerospace's Earth orbiters had simple fault-protection systems. Their satellites kept track of their own temperatures, voltages and processes. If an alarm bell went off, fault protection forced the spacecraft into an abrupt nap—safe mode—putting systems on standby, ordering attitude control to turn the solar panels to the sun, and instructing the communications system to send an SOS to the world. The spacecraft could orbit until folks on the ground solved the problem.

But going into safe mode could kill a mission if it happened during a critical sequence such as a planetary orbit insertion, when a few seconds determine whether a spacecraft flies beautifully around or crashes into a planet. A high-speed encounter with a comet 83 million miles from Earth was a big-time critical sequence. What if, for example, 10 minutes before impact with Tempel 1, the flyby's primary computer took a cosmic ray to the jaw in the middle of an orbit calculation? Mission controllers wouldn't get the news until three minutes before impact, and even assuming instant response, their help would arrive well after the comet roared past.

JPL liked fault protection. In 30 years it had become part of the lab's organizational DNA. Ball Aerospace understood the need, but harbored no visceral affinity for such fancy systems. They had built orbiters for 40 years, and orbiters did fine without fault protection. Alan Delamere, Bill Frazier and others had said from the start that the system JPL was designing for Deep Impact was too elaborate for a low-budget Discovery mission. But JPL stood firm, and the fault-protection system grew into the single greatest philosophical schism between the engineers at Ball and JPL. As the system evolved, Ball engineers worried more and more about it. And they worried about Kevin Barltrop, the JPL engineer who almost single-handedly shaped Deep Impact's version of the old Pathfinder/Deep Space 1 system.

The issue wasn't with Barltrop himself, who was widely considered brilliant, and a nice guy, too—calm, soft-spoken, a fine young engineer of 36 who had worked on Cassini's attitude control system. The man had studied Deep Impact's subsystems and their interactions, discerned the *de rigueur* from the dangerous, codified it all and tried to make sure the resulting fault-protection system worked.

What worried Ball was the scope and complexity of the system Barltrop was creating, and the fact that its intricacies became increasingly locked between his ears. The software tapped into every system on the spacecraft, with 49 "monitors" keeping tabs on the health of computers, amplifiers, sun sensors, power systems, temperatures, the gimbal, the navigation systems, the instruments, gyros, star trackers and so on. It compared what it found against 921 "symptoms," which in turn mapped to 667 "alarms," better known as faults. Depending on the fault, fault-protection software could launch one of 39 "responses" broken into 616 "response sequences" including as many as 150 commands each. Responses ranged from reissuing a command, resetting or rebooting hardware or software, switching to a backup or powering something off. As a last resort, the system could also punt like an orbiter and put Deep Impact into safe mode, which would power most everything down, turn the solar panels to the sun, and call Earth for help. Of course, going into safe mode at the wrong time—during a critical sequence, say—could kill the mission the system was intended to save. So fault protection was an important bit of software.

Its scope was formidable, its intricacies Byzantine. In a 2004 paper he co-wrote about his system, Barltrop described a key parameter: "E.g., GetUrgency(ValueSdst, Id, &InputSdstPrime, GetFault()) is evaluated at the very beginning of a response due to MonComTwta triggering SymComSdstRfLoss(B), mapped to FaultSdst(B), In this case, Id = B; assume &InputSdstPrime = [1 1], meaning Sdst(A) or Sdst(B) can assume primeness....."[1]

It was all too much for one human being to take on, but Barltrop became the expert no one else could help. Ninety-hour workweeks were routine for him, peaking at 120 hours on a couple of occasions, three standard workweeks

Kevin Barltrop

in seven days. He napped for a couple of hours at a time on a cot in his office, setting a cell-phone alarm to wake him to check on a test run. The janitor emptying his trash woke him often. He subsisted on Jack-in-the-Box fast food and ice cream. He grew pale even by Deep Impact standards. To a colleague who asked how he was doing, he answered, "Like a car running two years after the check engine light came on."

Management at Ball and JPL recognized Barltrop for what he was: a walking, talking, single point of failure. Indispensable and without backup, single points of failure were something space program managers avoided. Henderson feared that Kevin Barltrop, in a sleep-deprived haze, might accidentally step in front of a bus.

Barltrop called Bank's cell phone from Pasadena when he heard about the aborted test at Astrotech. He was angry. It was a legitimate test, he argued—why not let it run to conclusion and see what happened? Running it again would take hours. They were behind. Now they were more behind.

"That wasn't a test. That was an experiment," Bank shot back, his blood up as well. "We're not here to experiment. If you want to do experiments, do them on a simulator, not the spacecraft."

The rocky marriage of JPL and Ball had evolved into one of a reasonably happy married couple enjoying separate hobbies, vacations, and bedrooms. Such spats laid bare cultural differences largely buried as the mission wore on. Bank expected—needed, in fact—a deterministic system: if A, then B. Something predictable, something he could point to when Henderson or Keyur Patel or Rick Grammier asked for proof that the spacecraft would do its job. Barltrop's fault-protection system was impenetrable to Bank and his colleagues. Rather than "if A, then B," he saw it as, "If A, then we're not exactly sure what's going to happen, but when all is said and done it'll be all right."

The clash boiled down to Ball's tendency to quantify the risk of something going wrong, weighing that risk against the cost of fixing it and making a decision. Barltrop had designed a system intended to catch problems with scant odds of springing up. JPL had seen six Rangers and many other spacecraft killed or wounded by freak failures. As a result, JPL believed in Deep Impact's fault-protection system and felt it could well save the mission. Ball saw the very same software as a time bomb.

Deep Impact had to fly now or never fly at all. The team in Florida continued testing fault protection and launch sequences until they seemed reliable.

Grammier, Patel and Henderson watched schedules and budgets and prepared for a cluster of tense reviews to convince NASA that Deep Impact would succeed. In late November, Grammier delayed the launch from December 30 to sometime after January 8, 2005, to give mission testers at Astrotech another few days to run commands through the spacecraft. Deep Impact had to fly by January 28 or it would never catch its comet. The team worked through Christmas, capping what Henderson's children had come to call "The Year Without Dad." Then, for good measure, they worked New Year's Day.

A glitch with the rocket pushed launch back to January 12. The mission had a pair of one-second-long launch windows on a given day, easily scrubbed by an errant cloud or gust of wind. A heavy front could wipe out a week on the launch pad. Deep Impact had to be locked and loaded for at least two weeks, coiled and ready to strike off at a moment's notice.

In the marshes of Florida's Merritt Island, egrets and herons waded in flooded trenches along two-lane roads, sidestepping alligators. Estuaries and wetlands alternated with scrub and pine forests on land so flat the Atlantic Ocean hid behind coastal dunes only a few feet high.

Under cloudless skies on January 12, 2005, a Boeing Delta II rocket waited on Pad 17B, part of Cape Canaveral Air Force Station, which shared Merritt Island with NASA's Kennedy Space Center. The eighth Delta rocket had launched the first Orbiting Solar Observatory 43 years earlier; the 311th would carry Deep Impact to a much different place.

The rocket, the height of a 10-story building, looked too big to be going anywhere soon. But it was raring to blast the fruits of a six-year effort and more than $300 million—a dollar and change for every man, woman and child in America—into the path of a distant comet.

The Delta II's first stage included its own liquid-fuel motor plus nine strap-on solid rocket boosters. During its first minute of flight, it would unleash energy at a rate enough to power more than a million homes.

Earth, filled with molten metal, is heavy for its size, and what happens on Earth tends to stay on Earth. It's as if a toll collector at the outer limits of our planet's gravitational influence accepts payment only in extreme velocity: seven miles per second, about 25,000 miles per hour, being the minimum for passage. That's about 10 times the speed of a shotgun blast, or New York to Los Angeles in six minutes. It takes a precisely controlled, directed, sustained explosion to escape Earth's gravity.

The lower sections of the rocket's sky-blue body had frosted despite temperatures in the mid-70s. The cold came from the liquid oxygen tank within, filled an hour and fifteen minutes before launch with fluid chilled to minus-298 degrees. From that moment, the rocket seethed as the oxygen boiled off like water on a skillet. Higher, beneath mission logos emblazoned on the rocket's shaft, were about 10,000 gallons of kerosene, a highly refined version of the stuff young Frank C. and Edmund B. Ball's wooden-jacketed cans once held. The rocket would burn it at a rate of 37 gallons per second until it was gone.

Above the kerosene tank were the second-stage and third-stage engines, and then the justification for it all—a spacecraft tucked in its fairing with solar panels folded like the wings of a cocooned butterfly. Thicker than the rocket itself, the fairing afforded its occupant more space while lending the Delta II a false impression of top-heaviness. Deep Impact dozed in the dark, its systems standing by.

DEEP IMPACT'S CREATORS, staring at computer screens in control centers on both coasts, were wide awake, despite the exhausting three-month push to get the spacecraft to the point NASA would let them launch it. There was the slight but real risk of a failed launch, which would wipe out years of work and provide NASA with yet another black eye. Even a "smooth" launch risked being too rough for the spacecraft. NASA Launch Manager Omar Baez described it as "a real violent ride, going into space. It's not a ride in your dad's Cadillac. You have to make sure [the rocket] works and doesn't turn your spacecraft into a bucket of parts."[2]

Deep Impact had survived vibration and acoustic testing. Nonetheless, Rick Grammier, Monte Henderson and the other Deep Impact leaders manning consoles at the Kennedy Space Center knew their spacecraft had an elevated chance of ending up a bucket of parts. The gyro welds could, despite Tom Yarnell's best efforts to buffer them, fail. If they were going to fail, it would be today.

Grammier, in a pressed shirt and tie, sat before a bank of monitors. The monitors in the Kennedy Space Center's Launch Vehicle Data Center displayed views of Launch Complex 17, and data ranging from wind speed and temperature to the pressure of the liquid oxygen tank and the status of Deep Impact's systems. On the front wall hung dual 84-inch screens projecting images of the rocket, and above them four additional flat-panel displays with video feeds. In a different setting, it would have made for a great sports bar.

The NASA launch control room and its metaphorical red button—by 2005, the click of a mouse launched rockets—were elsewhere, but Grammier would give the Deep Impact team's final go for launch.

To Grammier's right sat Henderson. Henderson had not slept the night before. He leaned toward the monitors in the dimly lit room and stared at the screens as if his concentration could improve the odds of the launch's success. In a glassed-in VIP viewing area sat David Taylor, the Ball Aerospace CEO, who had notes for two very different speeches in the pockets of his jacket, depending on how the day unfolded. Taylor's boss, Ball Corporation CEO David Hoover, was hobnobbing with other VIPs a couple of miles away. The Muncie jar maker was now a Colorado can and plastic-bottle company, but Ball's leaders still took an interest in its side business in space.

With 20 minutes left on the countdown, Henderson said, "Flight segment is a go for launch."

STRONG WINDS AT HEIGHTS above jetliner cruising altitudes subsided. A malfunctioning tracking radar was behaving again. Ocean breezes were well below the 33-knot limit for launch. Things looked good.

Minutes passed as technicians exchanged acronyms and abbreviations on multiple audio channels. Launch teams reported final checks. It took a small army to orchestrate a rocket shot. So precise was the mission's choreography that if the spacecraft didn't launch at 1:47:08 p.m., it would have to wait another day.

Sixty seconds before launch, the spacecraft was unplugged from the tower. The flyby's battery could sustain it for 65 minutes without juice from the solar panels. The clock was ticking.

Outside, down miles of Florida's Space Coast, tens of thousands gathered on the beaches and piers to watch the Delta II launch. In his beach chair, Joe Galamback looked like another vacationer enjoying a sunny day at the ocean. With wife Donna and two teenage daughters, he was part of a group of 15 encamped across the bay from Cape Canaveral Air Force Station since late morning. His friend Alec Baldwin, who had, with Galamback, assembled Deep Impact with his own hands, was among them.

As Ball Aerospace's lead technician for Deep Impact, Galamback had been at the launch site from 2 a.m. until breakfast, tending to last-minute checks of the spacecraft. Now he listened to the countdown on a short-wave radio he had bought to follow NASA's play-by-play. His gaze was fixed on the Delta

II, visible across the water three miles away, past the Trident Wharf nuclear submarine docks where Galamback for months had fished for red snapper. The woman on his radio began counting backwards from ten.

Smoke consumed the rocket. Cheers and clapping nearly drowned out what she said next: "We have ignition and liftoff of a Delta II rocket carrying Deep Impact—NASA's journey to unlock the mystery of the solar system's origin."

The rocket rose, its fires like a small sun, until the din of Earth's canopy cracking and tearing stormed across the water. Galamback felt the sound go right through him, and it flushed him with adrenaline. For two minutes, he and his wife and their daughters watched the rocket until it disappeared over the eastern horizon.

No one said a word. Galamback thought: *It's gone. It's not coming back.*

Tears ran down his cheeks.

"I imagine it's like when your kids move away," he said later, trying to find words for the sudden, profound emptiness he felt.

THE DELTA II PERFORMED a supersonic striptease. Six solid rocket boosters burned out just over a minute into flight, backflipping away and plunging into the Atlantic. The remaining three boosters lit, burned and followed suit a minute after that. Its kerosene exhausted, the first stage separated two-and-a-half minutes later. Deep Impact was 70 miles above the Atlantic and already in space, traveling at 13,900 mph.

The second stage burned for five more minutes, increasing the craft's speed another 3,400 mph and lifting it to an altitude of 106 miles. The protective fairing split and tumbled off like a pair of giant tulip petals, giving Deep Impact its first glimpse of the sun. The second-stage rocket extinguished itself. The spacecraft coasted for 17 minutes to adjust its trajectory. Night fell quickly.

Twenty-seven minutes after launch, Deep Impact was over the coast of southern Africa. The second stage reignited and burned another 90 seconds before separating, and the third stage kicked in for a similar turn. The third stage released two giant yo-yos, which acted like a figure skater's arms extending to slow a spin, and then the spacecraft blew the explosive bolts connecting it to the rocket, setting itself free. The launch was going flawlessly, as reported and confirmed by tracking stations in Antigua, Ascension, Namibia, South Africa, and Diego Garcia in the Indian Ocean. Forty minutes after leaving Florida, it was time for Deep Impact to make its first call home.

•

The Deep Impact spacecraft atop its
Delta II rocket prior to launch

Deep Impact lifts off at 1:47:08 p.m.
EST on January 12, 2005.

The mission's nexus switched abruptly to Deep Impact's mission operations center at the Jet Propulsion Laboratory. In Pasadena, about 50 JPL and Ball Aerospace engineers watched the launch from a mission-operations area carved out of the eighth floor of Building 264. Their excitement about the successful launch was short-lived. They knew only that their spacecraft was traveling in roughly the right direction at 23,257 mph.

Communication with Deep Impact flowed through NASA's Deep Space Network. The network relied on massive dish antennas, each dozens of meters of across, in California, Spain and Australia, to converse with spacecraft. With Deep Impact scarcely out of orbit, the initial exchanges should have been loud and clear. For an instant, they were. But then came an extended, unnerving silence. Minutes ticked by.

Tom Bank, at mission control in Pasadena, considered what might be happening.

You launch, the fairing comes off, you get a little bit of sunshine, he was thinking. But it's doing a barbecue roll, so it's not getting much of a charge. The solar panels are still folded. So we're probably running in a degraded mode, spending battery power. That's bad. If the battery dies, we've got this great big, cold, frozen spacecraft up there with no hope.

Bank glanced at various JPL monitors. He put his feet up on a desk. He put them back on the floor. He was neither panicked nor relaxed. On one hand, he was an engineer who refused to jump to conclusions before the data was in. On the other, he was a human being who had worked five long years on this spacecraft.

"So here's your baby," Bank recalled. "It's just been born, and you're waiting for the first scream. And you don't hear it. Then you wonder—is the kid breathing?"

Finally, a signal. The spacecraft was communicating, but at a trickle of about a character per second. The kid was breathing.

Its fate depended on the sun. Perhaps the spacecraft was still spinning, its onboard thrusters having failed to stop its rotation after Deep Impact separated from the rocket's third stage. Or perhaps it was tumbling out of control, the cruelty of launch having broken the flawed gyroscope welds. In that case, Deep Impact could blurt out a bit of data when its antenna happened to aim toward Earth. But then only until the battery died.

Deep Impact communicated so slowly that mission engineers feared they wouldn't understand the problem until it was too late.

Finally, with what they estimated to be about 15 minutes of power left in the flyby's battery, they pieced together an explanation. The fault-protection software had "safed" Deep Impact because Barltrop's system had noted high temperatures in Deep Impact's thrusters.

To kickstart hydrazine combustion, thruster catalyst beds had to be warmed by some of John Valdez's heaters, usually to the 122-degree to 392-degree (150 C to 200 C) range. Among the many monitors in the fault-protection system was one keeping tabs on the lobe of the spacecraft computer controlling the heaters. If thrusters got too cold, the fault-protection system would heat them up.

Once a burn started, though, hydrazine stoked its own fire, cooking the catalyst beds as hot as 3,600 degrees Fahrenheit. Valdez designed his system to ignore anything over about 1,800 degrees during a thruster firing. Barltrop's fault-protection software pinged Valdez's system a few times. It decided the consistent 1,800-degree readings were a glitch with the temperature sensors and swapped over to backup hardware. The backup hardware interrupted firing thrusters. That caused the spacecraft's attitude control system to stumble in its effort to lock the solar panels onto the sun, which led the fault-protection system to tell Deep Impact to pull over, turn off the engine, turn on its hazard blinkers, and wait for help. A communication lapse between Valdez and Barltrop had led the fault-protection software to mistake perfectly normal behavior for a mortal threat. It was an example of the sorts of problems that make it into space despite years of testing.

Mission engineers responded quickly, telling the spacecraft to get back on the road. With what they thought to be about seven minutes of battery life left, Deep Impact sent word that its solar panels had opened and were having a good drink of sunlight.

The fault-protection software had pulled the plug late in the hydrazine burn to slow the craft's spin. As luck would have it, Deep Impact had ended up sailing with folded panels facing the sun. The battery had actually been charging. But the spacecraft just as easily could have had its solar arrays pointed at the stars in the opposite direction, which were far too distant to offer sustenance.

"The right way to think about it is we were this close to death," Bank said later, holding his finger and thumb narrowly apart. "And the Ball perspective was, we weren't that close to death. We were that close to shooting ourselves in the head, because there was nothing wrong with the spacecraft."

At least not that he was aware of. In three days, Deep Impact would take its first practice shots at a target far bigger and brighter than any comet. Bank's confidence would be put to the test.

Chapter 23

Rough Cruise

The moon was an obvious astronomical target, and one many on the Deep Impact team had been eager to capture. Unlike stars, Luna would fill the High Resolution Instrument frame, as the comet would, and provide a handy means of calibrating the sensors attached to the big flyby telescope and its smaller partner. Moving as fast as Deep Impact was, the lunar photography had to happen quickly.

JPL's decades of experience had taught its managers to wake up spacecraft gently, one system at a time, to give their fledglings a chance to sample life outside the nest slowly and deliberately. The scientists, on the other hand, were like kids with a new toy. They wanted to shoot for the moon.

There was a compromise. Four days into its journey, the spacecraft turned, the instruments snapped, the computer processed, the antenna spoke, and the Deep Space Network's 34-meter ears listened.

The smaller Medium Resolution Instrument's took in a sharp waxing crescent. The High Resolution Instrument, which should have captured crisper images yet, delivered a shock. It was clearly the moon, but with craters smoothed over like sand castles after the first waves of an incoming tide. The greatest telescope ever to fly beyond Earth orbit—the mission's scientific jewel—had delivered blurry images.

The moon as viewed by the High Resolution Instrument

Murk Bottema's COSTAR and the upgraded JPL camera may have reinvigorated the Hubble Space Telescope years ago, but the nightmare of the mirror flaw was still vivid in the space agency's collective memory. "You couldn't imagine how bad that is," Monte Henderson said. "NASA, they just can't handle blurry telescopes."

Henderson invoked the 48-Hour Rule. He asked NASA and JPL for something akin to an oath of silence: *There may be a problem. You now know about it. Give us two days.*

It would take much longer.

•

THE BIG TELESCOPE'S BLUR was one of many bumps on what was turning out to be a rough cruise. The fault-protection glitch had cost a day of moongazing, which scientists needed to calibrate the spacecraft's instruments, in particular the infrared spectrograph. Scientists knew what the moon looked like and what it was made of. They also knew how their flying spectrograph was *supposed* to perceive infrared bands of light. But the team couldn't be sure what their High Resolution Instrument's infrared spectrometer was *actually* seeing until they compared its spectral data to a standard reference. The moon, their standard reference, their tuning fork, had faded quickly out of sight. Stars and planets were too distant to finish the job. Doubt shadowed Deep Impact's spectral imagery.

ANOMALIES

1. Safing from catbeds at launch
2. MRI overtemp power-off
3. Attitude and rate spikes
4. Instrument image PDU drop-outs.
5. Pointing offset during calibrations
6. Star tracker 90° alignment error
7. Star tracker drop-outs during stare
8. Instrument bench too warm.
9. Anomaly List   too big

A JPL whiteboard eight days after the Deep Impact launch

The antenna gimbal acted up. Star trackers fed the data into a landscape format when the attitude control system expected portrait format (someone in attitude control had missed a detail in the star-tracker user's guide[1]). Such a mistake would have doomed the hard-wired Orbiting Solar Observatory. For the Deep Impact team, it was an easy software fix, beamed up the day after launch. Charged particles crashing in from a massive solar flare then blinded both star trackers, triggering a cascade of events leading the fault-protection system to plunge the spacecraft into safe mode again, this time justifiably. It took two days to get rolling again. Separately, fault protection shut down the flyby spacecraft's Medium Resolution Instrument for keeping its eyes open too long, despite a stray-light test having demanded it. Deep Impact was shaping up to be, as JPL mission engineers put it, a "tough bird to fly."

Flying the spacecraft was only part of what consumed the Deep Impact team. To get the spacecraft launched, Ball and JPL had set aside the mission's unfinished script. The mission operations team learned to fly Deep Impact while they calibrated instruments, tested thrusters, and checked out their autonomous navigation system using the moon and Jupiter and Saturn as targets. Meanwhile, a second group focused on fixing the encounter command sequences.

Rick Grammier walled them off and put his deputy Keyur Patel in charge of a team that included some of Deep Impact's top JPL and Ball engineers.

Jennifer Rocca

Steve Collins, a JPL engineer, works the Deep Impact test benches.

Grammier went as far as stripping most of their names from the mission organization chart. They had their own mandate, their own org chart. They were the Encounter Working Group, and they had one job: to make sure the last 36 hours of the mission *worked*. The choreography didn't work yet, and until it did work, Deep Impact stood a sizable chance of ending up a bullet missing another bullet not being watched by a third bullet. In the Encounter Working Group were Mike Hughes, Jeremy Stober, Tom Golden, Lew Kendall, Kevin Barltrop and a dozen others, including Jennifer Rocca.

Rocca had been testing Deep Impact since 2002 and had, in the intervening two-and-a-half years, worked the equivalent of a half decade. JPL bore much of the responsibility for developing and testing the mission scripts. It had been Rocca's job, and that of a few other JPL colleagues, to work with scientists on one hand and engineers at Ball and JPL on the other, then devise command sequences to both bring home the data and keep the spacecraft running smoothly.

She had suffered from John Marriott's decision to cut back on spending to refine the software test benches. There was no "SoftSim," standard on most JPL missions, which packs a low-fidelity version of a multi-million-dollar, multi-computer test bench onto a single PC. Software simulators let testers run early drafts of command sequences quickly, exposing big problems before taking hours on a test bench. Further, while the Deep Impact test benches simulated the universe and the spacecraft's systems, they were a nightmare to operate, forcing testers to write long strings of commands. JPL attitude control engineer Mike Hughes likened the simulators to the truck the Beverly Hillbillies rode into town.

"I mean these things were just smokin' and steamin' and kickin' and clankin'—they would break down…it was just a mess," Hughes said. There was a backlog to the point that "it was like six people with laundry lined up behind a washing machine."

Rocca insisted that the virtual spacecraft run through an entire imaginary launch before she was willing to simulate anything in space—to make sure all tests were run on the same playing field. That took four hours. Also, until about six months before launch, three of the four test benches had remained at Ball. The magic of the Internet notwithstanding, JPL testers lost half their runs to network problems, sometimes hours into a test.[2] They had no choice but to start over, further delaying a packed schedule and stretching the team's hours on both sides of the Rockies. Marriott's stinginess on test-bench spending had saved Deep Impact a mint, but only because Rocca and others at JPL and Ball were donating a fortune in time to the mission's cause.

Monte Henderson admitted that the test benches were "not great." But he felt JPL's expectations were outsized, assuming as a birthright Cassini-style test benches, immaculate machines, something "you could fly the *Enterprise* with."

"We had the budget for the Clampetts' car," he said.

Rocca had sensed the immense effort before her. Just 26 when she joined the mission, she sent an e-mail to friends. "I'm not going to see you for three years," she wrote. "I'm not being mean. But I'm honestly that busy with work that I can't."

Like Barltrop, she had a cot in her office, in her case one garnished with a Care Bears flannel pillowcase. Between her dual flat-screen monitors in JPL mission operations sat a Hasbro Care Bear named "Wish Bear." It was aqua, with hearts sewn into its feet and a shooting star embroidered on its little chest. There was a story behind it.

Rocca recalled that Barltrop, who had been working for two days straight, had penned into a test log: "I think there are little blue men in the test bed looking at me. They may have pressed some buttons and screwed up my test." Rocca decided a team working this hard needed a mascot. She did some quick research on the Web and found Wish Bear, which "helps make wishes come true, and although they don't always come true, making wishes and working hard to help make them come true is still a lot of fun."[3]

Perfect, she thought.

Rocca had grown up in South Fork, a town of 600 on the Rio Grande in southwestern Colorado. As a child she would wrap her Barbie dolls in aluminum foil and pretend they were on different planets. Rocca had long blonde hair and was cute, and when she had told her high-school guidance counselor she was thinking about going into aerospace, he said, "You know, Jennifer, aerospace is just not a field for women." She would flunk out after her first semester at the University of Colorado, he had warned.

Rocca got straight A's and sent him a copy of her transcript. Graduate school at Stanford followed, and at JPL she was soon known as intense, productive, and not afraid to raise hell for her causes. She was "extremely anal-retentive in everything she did" and "would come to Ball with a box of hand grenades," as a JPL colleague described it, and said it would take three people to replace her. "Her hair is on fire," added Patel.

Rocca was exhausted, but there would be no letup. The Encounter Working Group's demands were crushing. They ran test after test, smashing headlong into a problem, looking into it—which often involved bringing in their Ball counterparts—and starting over. They waited hours for the spacecraft to virtually launch itself for the umpteenth time, and then smashed into the next problem. They paid the broader implications of their work less and less mind, focusing instead on the next handhold or foothold, because, Rocca recalled, "Honestly, the mountain we had to climb was so high I think if we looked at it, we might have given up." And they had to climb quickly. The comet would not wait.

Early in a mission, many teams share responsibility for the ultimate fate of a spacecraft. Later, the field dwindles as accountability concentrates in fewer and fewer hands. At some point, like a relay baton, the success of the entire enterprise lands in one's palm—whether you want it or not. Rocca and her Encounter Working Group colleagues had grasped the baton at a very late hour. Their collective response, as Rocca put it, was to *get the fear* that if they failed, everyone would fail.

Jen Rocca's office cot

"I don't really know when the moment was, but at some point our team just got the fear, and it didn't matter whether you worked for JPL or Ball," Rocca said. "It just mattered that we needed to get it to work. And whatever it took, we just did it."

Rocca more than once worked three days straight, catnapping on her Care Bear pillow. She went home to do laundry, shower, and sleep two hours, and then drove the 12 minutes back to JPL. She worked through migraines and somehow drank no caffeine.

Morale sank. A JPL whiteboard filled with quotes. "So, who shoved a road flare up Keyur's butt?"; "Davey Crockett had better odds at the

Alamo"; "Deep Impact lesson #1: Never get good at something you hate"; "With enough ignorance, anything is possible"; "To realize the DI dream, we must first experience the nightmare"; "Note to self: Self, don't work on this project."

Offices and cubicles took on the look and stink of dorm rooms. People wore the same clothes, neglected personal hygiene, grew stubble. There was the impression that Deep Impact project managers and JPL encouraged self-neglect. As early as May 2004, more than a year before the planned encounter, Charles Elachi, the JPL director, stopped in for a JPL team meeting. Real help—in the form of user-friendly test benches, a SoftSim, or a doubling or tripling of mission-testing staff—was not in the cards. Someone mentioned it would be nice if real nourishment were made available. Everyone was living on vending-machine junk food.

A glass-doored appliance appeared. It had been the Mars Exploration Rover team's ice cream freezer. Deep Impact, as JPL attitude control engineer Steve Collins put it, was now like "trench warfare with free ice cream." Hughes described the experience as "like being in Bastone surrounded by snow in the engineering world." They were well-paid engineers, and nobody was trying to kill them, at least literally. But there seemed to be no way out.

Things were only somewhat better at Ball, where there were no cots, at least. But many still worked extreme hours. Marcy Kendall, whose marriage John Marriott had saved before it started, brought dinners back to the office for her husband Lew, the attitude control engineer. She later joked that the first year of marriage had been easy: she hardly saw Lew. Amy Walsh faced not only the demands implicit in having become the most vital person on the Ball side, but also the stress of her father's dying of cancer. She simply forged on. The finality of the date with the comet helped keep men and women going in Boulder as well as Pasadena. Independence Day would be their liberation.

Still, some in Boulder shook their heads at the JPL workloads and how hard that people like Rocca were pushing themselves. In Rocca's case, they saw it as the youthful exuberance of someone out to prove herself. She was taking on too much, not delegating enough. It all fit with the Ball impression that the JPL personality had a masochistic streak, a "hero paradigm" whereby mission success depended on superhuman efforts of the sort being extracted from Rocca and company.

Yet Rocca had more than one meeting with managers who said, "OK, if you work another 120-hour week, we're going to take your badge away," she recalled.

"I'm like, I'll give it to you right now. I'm not here because I'm enjoying it, you know? It's very simple. If you don't want me to work as many hours, give me less work to do," Rocca said.

They couldn't give her less work to do. *The comet would not wait.* The American taxpayer, represented by NASA Headquarters, could be asked to pay no more. The testers were on their own. The entire effort, from the mission's earliest designs to the laborious creation of hardware and software to all the scientific preparation—the work of hundreds, even thousands of people over several years—now came down to Rocca and colleagues at JPL and Ball ensuring that Deep Impact could follow its lines. It was either do the work of four people or let down hundreds of buddies. And they were all buddies here.

Rocca and teammates like Barltrop, attitude control lead Mike Hughes, and the navigator Dan Kubitschek became "like your family. These people…if I'm in a corner, they're going to back me," she said.

She thrived on "fighting through the challenges so you can experience the payoff at the other end," as she put it later. However, she added, "I don't think any of us realized the kind of sacrifice we were in for, or I would have said we were crazy."

THE HUBBLE SPACE TELESCOPE's focus had been distended and misshapen. As days turned into weeks, Deep Impact's focus appeared to be simply off target. The gap between the primary and the secondary mirrors had been set too wide. The point of focus, as far as engineers on the ground could tell, was a bit less than a centimeter in front of where it was supposed to be.[4]

What could cause such a problem? The imaging team, looking at the flawed pictures and running their computer models, calculated that the primary and secondary mirrors were about a tenth of a millimeter too far apart: the thickness of a sheet of copier paper. In the world of precision optics, it may as well have been a mile.

They hoped water was the problem. The big telescope's six-foot barrel was made of graphite. It had absorbed humidity during its time on the Florida launch pad. Perhaps it had sponged in more than expected. They would just "bake out" the moisture with the dozens of coaster-size heating pads Jim Badger and Tom Yarnell had glued around the length of the telescope's barrel years earlier.

The engineers' math told them it wouldn't solve the problem, and the math was right. The focal point improved perhaps a millimeter. A second, extended bake-out lasting from February into March won back another millimeter. By then, the barrel was dry. They were still seven millimeters away. The problem had to be somewhere else.

Jim Baer, Deep Impact's lead optical engineer, had had a sinking feeling from the moment he saw the big telescope's smoothed-out lunar images. He thought back to the testing regimen, a months-long affair involving a panoply of lasers and mirrors and equipment. There had been one step that had made him slightly uncomfortable.

Ball tested some of the world's finest space instruments in what looked like a beached submarine someone had painted sky-blue. It was a thermal-vacuum chamber named Rambo. Rambo, like the much larger Brutus around the corner, could suck the air out of its guts and then cook or freeze its victims at will. Optics designed to do their work in unearthly cold must be calibrated in unearthly cold. In sports, it was referred to as "practice as you play." In aerospace, it was "test as you fly." Instruments had to feel at home in Rambo's hell.

Rambo had a prominent porthole. Ball engineers used the window to fire lasers into precision optics in airless deep-freeze. The equipment calibrating the Deep Impact telescopes neither fit inside Rambo nor appreciated its harsh treatment. The window was also a problem. Windows bend light. To eyes as sharp as Deep Impact's big telescope, Rambo's window was like a pair of Coke-bottle eyeglasses. This two-inch-thick fused silica, on which Ball had spent $26,000 in the early days of Deep Impact, had to be precisely understood.

Baer and his colleagues came up with a creative solution. It was also quick and cheap. The telescope testing was going on in late 2002, just as the mission faced its most dire cancellation threat. They would use test hardware from the Spitzer Space Telescope project. Ball had built two of the three instruments and the liquid-helium-cooled heart of the infrared Great Observatory. Although the space telescope had yet to launch, Ball had delivered its share of Spitzer for testing at JPL earlier in the year. Among the Spitzer leftovers was a

The cryo-flat

mirror about the size of a dessert plate designed to remain absolutely flat in temperatures as cold as Rambo could muster. The idea was to put this "cryo-flat" into the chamber, have Rambo freeze it 100 degrees colder than dry ice, and bounce laser beams off it through the window. The cryo-flat being flat, the window would be the cause of any distortion. The engineers could then pull the cryo-flat aside and, understanding the window's idiosyncrasies, compensate for them when focusing the telescopes.

But what if the cryo-flat wasn't…*flat?* What if the mirror—and not the 18-inch porthole into Rambo—redirected the calibrating lasers in unexpected ways? Such quirks would literally be built into the Deep Impact's High Resolution Instrument—and its lesser telescopes, too, though they would be too small to notice.

Confirming the cryo-flat's actual flatness in deep-frozen vacuum would have taken equipment Ball didn't own, money the mission couldn't spend, and time the instrument team didn't feel it had. It took days just to pump the air out of Rambo and chill its innards to minus-220 degrees.

Deep Impact's blind faith in the cryo-flat had nagged at Baer. He considered buying a $3,000 flat to compare with the Spitzer hand-me-down, but he would have had to tape the new mirror to hardware inside Rambo and test the porthole in parallel with other scheduled deep-freeze tests to avoid spending the few thousand dollars a day it cost just to run the vacuum chamber. His gut told him he should do it; his brain argued otherwise.

His brain made a good case. It might not work anyway, after all. His colleagues might not take well to his taping up their precision optical fixtures. Even a few thousand dollars, petty cash in the aerospace business, was hard to come by amid Deep Impact's financial woes. Plus it was a combination of crunch time and Christmas. So Baer and the instrument team just went with the Spitzer cryo-flat.

A'Hearn, who had taken over the instruments' management from JPL to save money, had been thinking about Hubble. Perkin-Elmer had done two different tests on the Hubble mirror. One test had said the mirror was fine; the other said it wasn't. Perkin-Elmer, under enormous pressure from NASA managers, had chosen to believe the good news.

For Deep Impact, A'Hearn had insisted on two independent focus tests that agreed, and Ball had delivered. Both tests, however, had been united in disagreement with where Tom Yarnell's spreadsheet model had argued the focus should be.

Ball's instrument team had decided to believe the hard data the test had delivered, and A'Hearn had concurred. He had thought the tests were so different as to be immune from what he called "some common-mode failure."

"It turns out there was one piece in common that killed us," A'Hearn said.

The cryo-flat. Corning had made it layer by layer, and those layers varied just enough that they contracted with the cold in vanishingly different ways. The cryo-flat became a cryo-concave. Its curvature was so slight that, if extended until the unwanted arc completed an imaginary circle, it would have had the circumference of Lake Tahoe. But it was enough to shift the focal point of a six-foot telescope about a quarter inch. Like Hubble before it, a blurry NASA space telescope had been painstakingly calibrated to precise imperfection.

On March 25, 2005, the project went public with the problem. There was no press conference of the sort Ed Weiler starred in 15 years earlier, but rather a press release titled: "NASA RELEASES DEEP IMPACT MISSION STATUS REPORT."

NASA began with the excellent performance of the navigation system and the superb accuracy with which Deep Impact was tracking comet Tempel 1. Then, in paragraph six, the mission mentioned its headline instrument having "not reached perfect focus."

NASA had learned from the Hubble debacle. Weiler, not shy about promising a fix, had spoken of technological weaknesses and human disappointment in his typical frank style. While Hubble's flaw had been technical, it became foremost a public-relations disaster.

The Deep Impact press release sought to preempt any such possibility. Attributed to Grammier was the statement, "This in no way will affect our ability to impact the comet on July 4. Everyone on the science and engineering teams is getting very excited and looking forward to the encounter." A'Hearn officially added, "It appears our infrared spectrometer is performing spectacularly, and even if the spatial resolution of the High Resolution Instrument remains at present levels, we still expect to obtain the best, most detailed pictures of a comet ever taken."[5]

This was true, though there hadn't been much competition. The infrared spectrometer's resolution was coarse enough that the blur didn't matter. But the focus error diminished Deep Impact's High Resolution Instrument to the equivalent of another Medium Resolution Instrument, incapable of resolving details of the crater the impactor spacecraft would leave behind. Jay Melosh, the cratering physicist perhaps most interested in seeing the divot Deep Impact would gouge, broke protocol. "I would say it's a major concern. We're all very disappointed at the performance of this imaging system," he told the *Rocky Mountain News*.[6]

The bad news, innocuously labeled, buried and timed for a Friday afternoon—a hectic time in newsrooms cooking up the bulk of Saturday, Sunday and Monday editions—slipped by largely unnoticed. The *New York Times* and *Philadelphia Enquirer* published briefs. Henderson's nightmare headline, "NASA Builds Yet Another Blurry Telescope," never ran.

·

HAD MURK BOTTEMA's COSTAR eyeglasses and the new JPL camera failed the Hubble Space Telescope, NASA had a Plan B. Scientists aimed to solve the space telescope's problems with computer software even before Ball began working on COSTAR.

Computer-based image reconstruction—powered by something called deconvolution algorithms—would clean up Hubble imagery. In theory, the software worked by calculating where the light of a perfect telescope would focus, then rearranging the mess of digitized photons coming in from an imperfect telescope.

Deconvolution software had existed years before Hubble. Intelligence experts used it to hone images of foreign military bases and missile silos, or to map out terrain with enough accuracy that a low-flying cruise missile would know when to bob or weave. Radio astronomers applied similar code to their cosmic images, and Voyager's famous photos of the outer planets in the 1980s had also enjoyed the touch of the digital airbrush.[7]

Hubble scientists had refined their mathematical equations over time, striking a balance between enhancing the actual images and amplifying strewn light to the point it might be mistaken for some amazing new astronomical find. They greatly advanced the science of fixing digital telescope imagery. Even before COSTAR flew, computer software had improved the sharpness of Hubble's vision to the point that the telescope could discern objects nearly as well as originally hoped.

The achievement came at a cost. Hubble's flawed mirror scrambled incoming photons into an optical tangle too knotted for a digital comb. The software had to throw away 85 percent of the incoming light, and with it a great deal of the sensitivity the telescope needed to see into the deepest recesses of cosmic history.[8] Sensitivity had been one of Hubble's big scientific selling points. The great contributions of COSTAR and JPL's Wide Field and Planetary Camera 2 had been to recapture most of the blur-scattered light in its natural, analog form—something software could never do.

On April 14, 2005, A'Hearn called a meeting in a conference room down the hall from his office. He wanted to understand how deconvolution software might clear up the Deep Impact telescope's vision. About 20 scientists and engineers attended, with others dialed in.

The good news, as with Hubble, was that the blur's cause was no mystery. Also, Deep Impact's key images would be of a bright target—a wounded comet and its trail of glowing innards—against the black backdrop of space. High contrast would ease efforts to digitally remaster images on the ground.

Deep Impact's straightforward problem of misplaced focus also lent itself to simpler rearrangement than Hubble's sloppier spherical aberration.

A'Hearn hired Ivo Busko and Don Lindler, who had helped develop deconvolution algorithms for Hubble. They created a special prescription for Deep Impact's digital eyeglasses, based in part on mathematical equations designed for the great space telescope. With their software, the two image mechanics could break down and rebuild a photo in a few hours, and with such fidelity as to almost completely restore the vision of Deep Impact's big telescope, scientists believed. Proof would come in July.

Chapter 24

Separation

Deep Impact's engineers, technicians, managers and staff had worked six years to craft two flexible, intelligent spacecraft out of materials forged in the Big Bang or in the bowels of stars, frozen in comets or baked into asteroids or cooked and churned in their rocky planet. They aimed to find out where the stuff of their bodies and their soaring products came from, how Earth and its solar system formed, and whether the wispy hazes around distant stars might also be cradles of life.

The team had relied on elaborate processes, meetings, spreadsheets, timelines, as well as diverse experience and technical expertise. To get it all working had taken determination, dedication, endurance, sacrifice and passion. Yet in the end, Deep Impact boiled down to six words: Find comet. Hit comet. Watch comet.

Navigators had found the comet. On April 25, 2005, the big telescope sent home a postcard of a white dot 40 million miles from its camera. It wasn't the comet itself, but rather its halo coma of gas and dust thousands of times larger, reflecting just enough sunlight for the High Resolution Instrument to perceive it.

Next was hitting the comet.

As the flyby spacecraft had warmed up for the big event, the impactor had mostly slept. Mission leaders hadn't planned on waking the smaller craft until two weeks before impact. But one evening, during yet another 80-hour work week, Lew Kendall, the Ball attitude control engineer, noticed something odd while wading through the seismograph-like plots of a test-bench run.

The impactor's lone star tracker—the attitude control system's eyes— seemed to be shifting the positions of stars in the corners of its field of view. He called his JPL counterpart Mike Hughes and suggested activating the impactor early to see if the same thing happened in space.

Hughes agreed, the impactor awoke, and the real star tracker indeed had the same problem. That wasn't all, Hughes found to his alarm. During the last weekend in May, less than six weeks before encounter, the impactor's star tracker recognized five stars to be among those in its catalog. But in fact, it was looking at eight stars. Three—distant, faint and absent from the planetarium show running through its brain—appeared so close to mapped stars that the star tracker mistook the pairs of stars for single ones. This misapprehension shifted the star tracker's estimate of its position less than two tenths of a degree, but it was enough to fool the gyroscopes into thinking the impactor spacecraft was slowly rotating.

Hughes sat in his home office, feeding his three-week-old son a bottle. It was Memorial Day. He considered the squiggles of his attitude control graphs. His eyes widened. "This is a grenade," he murmured. If the star trackers grew this erratic in the last 24 hours of its flight, AutoNav, the autonomous navigation software, would interpret the phantom rotation as the spacecraft being off course—even if it were right on target. AutoNav would command the thrusters to fire and push the craft out of the comet's path.

Missing the comet would render Deep Impact's true imperative—watching the comet—among the most anticlimactic moments in the history of spaceflight. The flyby would still shoot the best-ever images of a comet, of course. But the errant impactor would deliver a Fourth of July dud, depriving NASA of its cosmic explosion, scientists of their pristine comet innards, and subjecting JPL and Ball Aerospace to public ridicule for letting a software error spoil yet another mission.

Word traveled quickly at JPL. There were rumblings that attitude control was going to blow Deep Impact. Hughes took it personally. He hung the Rudyard Kipling poem "If" on his cubicle wall. It had fueled his fire at Penn State. *If you can trust yourself when all men doubt you/But make allowance for their doubting too…If you can force your heart and nerve and sinew/To serve your turn long after they are gone…Yours is the Earth and everything that's in it/And—which is more—you'll be a Man my son![1]*

Hughes and Charlie Schira, the attitude control leads at JPL and Ball, explained the problem to the JPL leaders Rick Grammier and Keyur Patel and to their Ball counterpart Monte Henderson, who sanctioned another tiger team. The group designed software fixes that worked on the test benches. One changed the way star trackers processed images, in particular ignoring the near-neighbor stars confusing the device; the other was to adjust how attitude control software evaluated star-tracker outputs before feeding them to the gyroscopes. Hughes felt good about the changes. Now he and his colleagues just needed Deep Impact to roll over in space for 24 hours. During that time, the impactor's star tracker could have a good preview of the stars it would see once the impactor released.

But the spacecraft was booked solid. Attitude control could have a few hours, maybe, but a whole day, no way. Deep Impact was busy with imaging for scientists, software and sequence uploads, science calibration, operational readiness testing, and, for the week until impact, continuous imaging of Tempel 1. Jennifer Rocca, the JPL engineer helping run the test-bench effort, had driven herself to the point she felt the need to ask Tom Golden, the Ball software engineer working only slightly fewer hours, to stand behind her and watch her type.

Rocca was about to send into deep space a mammoth sequence to guide the paired spacecraft through most of its final week. There had been no time to run it all on a test bench. Dangerous, Rocca knew, but the mission would have time to recover should something go wrong. She needed another set of eyes on her screen.

"Why?" Golden asked.

"I'm making edits to this," she said, referring to the sequence.

"O.K. But I don't know what edits you're making."

She was so sleep-starved her hands trembled. "Well, if you could just watch me so I don't make a typo? That would be great."

And away the script went, leaving attitude control engineers Hughes, Kendall, Schira and the rest of the Deep Impact team to wonder whether their copper bullet would find its mark or run from it.

SURPRISES IN ONE PLACE brought scrambling in another. Back on Earth, deconvolution software could sharpen the blurred images coming back from the big telescope, maybe, but that did little for autonomous navigation software demanding real-time views of the comet on approach. In particular, the High Resolution Instrument was to be the flyby's eyes for scene analysis, Nick Mastrodemos's software for choosing the bright spot to hit. JPL engineers scrambled to train the flyby's smaller Medium Resolution Instrument as a stand-in.

Navigators also noticed ephemeral star-like blots in some of the images coming back from the cruising spacecraft. These were cosmic ray hits—heavy ions such as iron, formed when distant stars exploded, smashing instrument detectors at close to the speed of light. There was no stopping them, and their remnants stood a chance of looking just like a comet to AutoNav until the spacecraft was perhaps four hours away. So in addition to all the scheduled testing, navigators launched into a crash program to simulate cosmic ray hits, modifying their software to ignore potentially distracting flashes appearing at the fringes of the telescopes' vision.[2]

Deep Impact had four major contingency plans—elaborate responses to worst-case scenarios—each of which had to undergo the same intense scrutiny and test-bench runs as the best-case scenarios. One would attempt to hit a comet shaped like a dumbbell or a donut or some other odd form. Another dealt with recovering if the flyby decided to switch over to its backup computer during the encounter's key moments. A third concerned an unresponsive impactor. In that case, the flyby would hang onto its dead partner 14 hours longer than otherwise planned, releasing it 10 hours prior to encounter. From the closer distance—roughly that separating the Earth and the moon—the dead impactor could be relied upon to meet its mark, although perhaps not in a sunlit spot. The flyby then would drain most of its remaining fuel in a mad sprint from the onrushing Tempel 1.

The fourth, and most poetic, of the plans, was called "Flight System Impact Mission." Here, the impactor, as if sensing its fate, clung to its larger companion, the flyby spacecraft. The flyby would try to convince its partner to let go, then give up and steer itself and the impactor into Tempel 1, snapping photos until the comet barreled over them both. Space telescopes—Hubble, Spitzer, Chandra, Swift, the Submillimeter Wave Astronomy Satellite—and ground-based observatories would have to collect what they could from 83 million miles away. Jeremy Stober, the Ball engineer who had helped design the impactor spacecraft and led the team building it, explored this scenario.

It was the Live for the Moment strategy taken to its extreme, involving entirely different communications, navigation, imaging sequences, and computer processing. Stober, running and rerunning Deep Impact through its virtual double suicide on a test bench in Boulder, worked for 36 hours straight to get it done. Two days later, in mid-June, Stober and 21 Ball colleagues flew to JPL.

On June 22, a blast of water and gas erupted from the comet, doubling Tempel 1's typical output. The outburst, triggered by the sun's heat, thrilled scientists, who pored over their spectral data. They created a short movie of the big telescope's photos of the event, posted it to the project Web site, and had a press release written up.

The engineers responsible for the craft's guidance had a completely different reaction. AutoNav could mistake the outburst for the comet itself and shift its target point off the edge of the nucleus, Hughes feared. The impactor might whisk through it like an airplane through a cloud.

But the scientists in general and Mike Belton in particular helped assuage such fears. Deep Impact's founding father was still on the science team. Belton had taken a particular interest in understanding the comet's rotation and had figured it to be about 41 hours. A smaller outburst on June 14 had come from the same place on the comet, he reasoned. It was probably a spot where the baked crust had cracked to expose volatile ices prone to jetting off in the sun's heat. Belton figured the fissure would have slipped back into the cold shadows by the time of impact, its flow stanched, the resulting halo dissipated.

BELTON AND HIS DOZEN science-team colleagues had played quiet but vital roles in the mission during the years of engineering and spacecraft construction. Besides perhaps Mike A'Hearn himself, JPL's Ken Klaasen was the most engaged with the mission's engineering team, serving as the link between the science team's desires and the mission scripts guiding the spacecraft. He helped orchestrate imaging sequences and figured out the orbit of the Hubble Space Telescope for navigators who could slow the spacecraft to ensure the greatest eyes in space had a clear view of impact. But the others also prepared for the encounter with Tempel 1.

Cornell scientist Peter Thomas wrote software to create a three-dimensional model of the comet based on two-dimensional images Deep Impact would shoot from various angles. His colleague Joe Veverka updated the science data handling system designed for his ill-fated Contour comet mission for Deep Impact. Don Yeomans of JPL continued to refine estimates of the comet's location in space. Mike A'Hearn kept the big picture in mind while making sure JPL and Ball delivered a mission capable of achieving it.

The youngest member of the science team, Jessica Sunshine, prepared for the spectroscopic deluge. Sunshine, in her late thirties, had impressed Belton with work she did for the Galileo mission to Jupiter. She was charged with translating the curves coming back from Deep Impact's infrared spectrometer into definitive statements about what was or wasn't in the comet. It would be a monumental task.

Sunshine prepared models to predict what various chemicals previously spotted in comets—methane, ethane, acetylene, carbon dioxide, formaldehyde, ammonia, water and 46 others—might look like coming back as spectral data from Deep Impact. The instrument would deliver information across hundreds of shades of infrared, or wavelengths, with many chemicals revealing themselves at multiple wavelengths. Rather than just looking at emissions from a hard surface, like the moon, or how a cloud of space dust absorbs light, Sunshine would have to contend with sunlight from the comet's surface and the surrounding dust, plus the gases wafting off the comet. Then, once the impactor dug in, there would be the glare from the comet's spilled guts. Spectral clues would be draped over one another like carpet samples, and she would have to somehow peel back those layers to make sense of it all—with a poorly calibrated instrument, no less. [3]

"A comet is a frustrating place to do spectroscopy," Sunshine concluded.

Karen Meech, a University of Hawaii astronomer, was to learn as much as possible about the comet from ground telescopes. To that end, she orchestrated the greatest organized astronomical campaign ever.

Deep Impact, traveling light, had thin spectral coverage and had to rely on distant help from more powerful instruments. Meech made a list of the world's great telescopes and considered their strengths. She thought about which telescopes should look at particular types of light—infrared, visible, radio and otherwise. Meech ran Deep Impact workshops for astronomers in Australia, Chile and Taiwan; the Taiwanese in turn helped Meech line up observers "in places I didn't realize there were telescopes" such as the hinterlands of Uzbekistan, China and the Himalayas.

Meech set up a computer system so observers could nail down who would be looking at what wavelengths with what filters and when, keeping in mind that the ground-based telescopes enjoying darkness at impact would roll into daylight while the comet still spewed gas and dust, and vice-versa. By the time Meech was done, some 80 large observatories had signed on officially, with many smaller institutions taking part informally. Every sizable telescope on Earth was going to be watching Tempel 1 at some point before, during and after impact, thanks to her.

Then there was the science of the crater itself. Just as snowballs on Earth varied in texture from skull-breaking to powdery, the true nature of comets had been a topic of dispute, even among the specialists on the Deep Impact team. Some believed the impactor would disappear into the comet like a sugar cube through cappuccino foam. Others speculated that rock-hard ice would yield little to the copper gnat. The science team had a friendly betting pool going as to how big the crater would be. Jay Melosh and Peter Schultz were favored to win.

Melosh, the University of Arizona planetary scientist, was a theorist. He made craters using math and computers running "hydrocode" numerical simulations originally developed to model nuclear explosions. Schultz, a professor at Brown University, blew real holes in real material with the fastest gun in the west. They were as different in personality as in approach.

Melosh was an easterner turned westerner, born in New Jersey, serious and headstrong. Schultz was a heartlander turned easterner, from Illinois, a senior professor who had been on Jessica Sunshine's PhD committee but somehow had managed to retain the enthusiasm of the child blowing up ant hills with mail-order gunpowder, which he once was.

Schultz spent two weeks a year at the NASA Ames Vertical Gun Range, where he served as science coordinator. Schultz and the range were both there because of the moon. His fascination with cratering was sparked when, as a child, he studied the moon's pocked face through a telescope eyepiece; his PhD dissertation became a book called *Moon Morphology*. The Ames vertical gun had been built for the Apollo missions. Simulating moon craters could tell NASA something about the body's consistency, so the agency could be confident the familiar gray world wouldn't simply swallow up the Eagle and its astronauts. More recently, scientists had used the big gun to better understand the wallop and aftermath of a comet or asteroid collision with Earth.

In his Deep Impact work, Schultz used the gun to shoot one-gram Pyrex spheres at up to 6 kilometers per second, or about 13,500 mph. Instruments including photometers, spectrographs, and a high-speed camera (loaded with a 400-foot roll of film, it could rip through 10,000 frames per second, enough for 7 minutes of a standard movie) captured the action. During week-long visits over five years, he and graduate students fired about 200 shots into everything from buckets of sand, pulverized pumice coated with red paint powder, garden perlite, and a sort of frozen pizza to which he added green food coloring, sugar and a dash of Worcestershire sauce as organic spice.

The perlite had been his wife's idea. He was looking for something to simulate a highly porous comet. The bubbly volcanic glass had done wonders for her garden and was known as a key ingredient in everything from lightweight ceiling tiles to sewage treatment. Maybe it could help humans understand the nature of celestial bodies, too.

The theorist Melosh was skeptical of the experimentalist Schultz's efforts. The physics of Deep Impact rendered even the Ames cannon a relative popgun, and the comet's gravity was hundreds of times weaker than what Schultz's frozen pizza had felt. The comet could be a block of cement or a marshmallow. Computer modeling would be the more fruitful road, he believed.

Schultz, for his part, felt there weren't enough hydrocode geniuses alive to crank out simulations diverse enough to span cement to marshmallows. One had to harness the complexity of nature, manifest in experiments, and do the best one could applying it to a related, if much larger, impact.

"We approach things very differently. Jay tends to work from first-principle physics. He's a brilliant guy and that's how he sees things, and he has strong opinions," Schultz said. "I tend to see things in alternatives. I have insights in experiments. Jay has insights from his theoretical works."

A'Hearn added, "They just very rarely agreed with each other. They tended not to talk to each other. They tended to talk to each other through me."

The point of both Schultz's and Melosh's work was ultimately forensic. Based on material blasted from Tempel 1 and the crater left behind, how hard or soft was the comet? How packed or empty? How heavy or light? From that, combined with Sunshine's spectroscopic insights, scientists might have enough hard data to bolster or toss out theories of comet and, by extension, solar system formation.

Melosh ran his simulations. Schultz shot his concoctions from different angles and found, for example, that perlite soaked up the blow with little cratering when struck from straight overhead, but yielded a sizable crater if the massive gun's Pyrex projectiles hammered it from a steep angle—the opposite of what one would find with an impact on sand or dirt or stone. He and his graduate students worked on mathematical models scaling the effects up to the higher speeds and greater forces Deep Impact would deliver. By the time they were done, Schultz hoped he could look at post-impact images and recognize explosive patterns he had already seen, generated through physics he understood. Now the engineers just had to deliver on what Schultz called "the grand experiment."

DEEP IMPACT WAS A GAME of space golf. The Delta II played driver. The flyby's five trajectory corrections during the long cruise served as long irons. The last chip shot, programmed into the sequence the jittering Rocca had sent with Golden's bewildered oversight some days earlier, happened on July 2. Thirty hours before encounter, thrusters pushed the craft toward a point navigators estimated to be within a kilometer of an imaginary bullseye on the comet nucleus, which was still too far away for the big telescope to see. Even if Tempel 1 looked like a banana, scientists figured its waist was at least two-and-a-half miles wide. If the spacecraft did nothing else, it would probably hit. The ball was rolling straight for the cup.

The comet was still 690,000 miles away, closing the gap by the width of Kansas with each passing minute. Deep Impact had finally mastered its lines, with all the contingency scripts in its back pocket. The Encounter Working Group's hellish half-year was over. There would be no more testing, no more surprises, analysis, configuration, testing, uploading or re-uploading. The team could just watch the numbers trickling back from the spacecraft and, if something really bad happened—a major anomaly, in Deep Impact parlance—pull the main script out of their spacecraft's hands and tell it to read from a contingency plan, such as Stober's "Flight System Impact Mission." There could be no "joy sticking," or manual control, anyway. Thousands of binary commands, each having taken days or weeks to formulate, test and retest, were now masters of Deep Impact.

Therefore, for much of the Deep Impact team, Saturday, July 2, 2005—a typically humid, sunny, temperate summer day near the southern California coast—held a strange mix of anticipation, anxiety and boredom.

Monte Henderson was talking to the press so much that Ball's media relations people joked he should sell his own Deep Impact Project Manager action figure, complete with utility belt stocked with a Blackberry, cell phone, laser pointer, and Tylenol PM. He had done TV interviews for crews from the national broadcasters of Japan, England and Australia, and spoken with dozens of print journalists in Pasadena for the big event.

Now, with the spacecraft's critical sequence rolling 82 million miles away, Henderson sat in the shade outside the Theodore von Karman auditorium. In a half-circle around him were print reporters from newsrooms as diverse as the Associated Press, the *Rocky Mountain News*, *Space News* and *Xinhua*. It was Henderson as teacher with the arrayed media as pupils.

The curtain had parted, and JPL, Ball, and Mike A'Hearn and his science team stood at center stage. The long road to this point was ignored, as it was with every mission—the near-cancellations, the cracked mirrors and gyro welds, the computers flying first class, the endless frustrations of mission testing . . . *his and his colleagues' lives for more than five years*. Now it was about the interplanetary skeet shoot: bullets hitting bullets watching bullets, or pennies pitched in front of speeding tractor trailers, or mosquitoes pancaked against Boeing 767 airliners, as NASA press releases put it—speed, power, explosions, the stuff of life. NASA had set the tone in early June, reminding the world's media outlets that July 4 would be the "Spectacular Day of the Comet," complete with a "first of its kind, hyper-speed impact between space-borne iceberg and copper-fortified probe . . ."

"We are really threading the needle with this one," Rick Grammier was quoted as saying. "In our quest of a great scientific payoff, we are attempting something never done before at speeds and distances that are truly out of this world."[4]

The press was fawning. Headlines across the country trumpeted impending celestial fireworks, meetings with dirty snowballs, cosmic collisions, comet bombings. Reporters and the public had little capacity to comprehend what it actually took to get here, where Deep Impact, without fear of cancellation, could cruise along through its critical sequence and actually have a shot at doing what Alan Delamere and Mike Belton and the others had dreamed up more than 10 years before.

Over-the-counter sleep helpers were to thank for what little rest Henderson had been getting. He looked exhausted. And he was. A reporter asked what the Deep Impact team had to worry about, given that the mission was on autopilot from here on out anyway.

"We have been working on this program for five-and-a-half years, yet all of the major milestones are still in front of us," he answered. "I feel like our spacecraft is in very good shape. Everything we've tested so far shows it's ready to do the job. But there's still the uncertainty of everything that's going to happen in a very tight, well-choreographed ballet in those final 24 hours."

He paused for a moment.

"It's the what-ifs. What haven't we thought about? There's so many things that we can walk through again and again and again. And we're just never going to be absolutely confident that this is going to work."

SEPARATION WOULD OCCUR precisely 24 hours before encounter. It was, to Deep Impact's engineers, the mission's most crucial moment. Tom Bank explained it this way.

"You're a billionaire," he began. "You decide you want to race in the Indy 500. You rent a garage right next to the track, you hire the best people you can, and you build the absolute best car you can. And when the green flag goes down, you start the engine for the first time, you open the garage door and out you go. You hit the first corner at 200 miles an hour. You haven't even touched the brakes."

The impactor had never flown except in simulation. Commissioning the flyby had taken more than two months, a typical interval to warm up systems and work out kinks. The impactor would have two hours. Should a major problem crop up, there would be no time to fix it. The little spacecraft had been tested to the hilt. But, Bank said, "There are so many things that could have been wrong in that testing. You do your best to test it. But testing is never perfect. And there are all these things that happen that you didn't think to test for. Because you're in a clean room. You're not out there in space."

At its most basic level, separation involved cutting the electrical cord—a 162-pin connector with heritage dating back to the 1970s-era Viking Mars program—and then breaking free mechanically with springs and pyrotechnic separation nuts. Ball had tested it all in Boulder and it seemed to work fine. They had gone as far as assigning an engineer named Bill Schade to create computer models demonstrating how the six springs pushing the impactor away would act once the bolts holding the spacecraft together let loose. What if one spring pushed very slightly harder? What if one bolt released a millisecond later than the others?

A sloppy push, Schade figured, could propel the impactor more than six miles off course, making the autonomous navigation software's job that much harder.

But that was only part of what was causing the team's collective separation angst. The impactor's battery might be dead, having somehow drained itself during its 172 days in space. Engineers knew only that it had been charged in Florida and that temperatures inside the impactor had been favorable to retaining that charge. The propulsion system's thrusters had never been fired. The radios on the two spacecraft hadn't conversed at distances more than a few feet. And the flyby had no experience flying without the impactor in its belly, despite all the miles it had covered. The impactor's attitude control system had its last-minute software patches. Who knew how the spacecraft's systems would perform together once unleashed?

Then there was the worst case: an impactor pronounced dead after release. The flyby would have already smoked through 30 kilograms of hydrazine, in essence slamming on the brakes for 14 minutes to avoid the frozen freight train. The mother ship could not turn back and crash into the comet itself. Should the impactor break down en route, the flyby craft would assiduously scan, image and send reams of data about a spot on a comet it deemed perfect for impact but at which nothing happened.

In southern California, separation would unfold after dark. The timing, 11:07 p.m. Pacific Daylight Time, was a function of the scheduled impact the following night. The science team established that timing, in turn, as a compromise involving a number of factors: the positions of Hubble and other space telescopes; major Pacific, American and South American ground telescopes; and a desire for overlapping Deep Space Network communications in the Mojave Desert and Australia at the moment of impact. NASA would have its July 4 fireworks, but only just—on the East Coast, the cosmic explosion would happen at 1:52 a.m.

NIGHT FELL, cool and pleasant, as if to remind visitors why people endured the traffic of greater Los Angeles. The engineers, manning mission control consoles 24/7, focused on the data coming back through the Deep Space Network. Nearly as far away as the sun, though more in the direction of Pluto, the spacecraft's reaction wheels spun up and the ship leaned back 45 degrees, diverting its telescopes to random star fields—taking its eye off the ball, as Delamere had so stridently argued against—to line the impactor up as precisely as possible with its distant target. Then, reading from its script, the flyby cut its umbilical ties, blew its separation nuts and nudged the impactor into the void.

Engineers in mission control stared at their screens. The impactor was alive, right on course and conversing with its partner, which, by the time the message reached their terminals, was poised to fire its thrusters to flee the comet's path.

Tom Bank sat next to Jeremy Stober as the data rolled in. He glanced at his young colleague's screen. The impactor seemed to be operating perfectly. Its eyes were sharp and its inner ears steady. The attitude control fixes had worked.

"We've got a chance of hitting this thing," Bank said aloud. "In fact, there's no way we can miss."

Stober turned to Bank. He was thinking the same thing. But he said: "Don't say that. Don't jinx us."

Fifty-one minutes later, the flyby spacecraft finished its escape, trained its eyes back toward Tempel 1, and snapped a photo. In the center of the image, blocking the telescope's view of the comet, was a delta-shaped star many times brighter than everything around it. It was the impactor, waving goodbye with a glint of reflected sunshine.

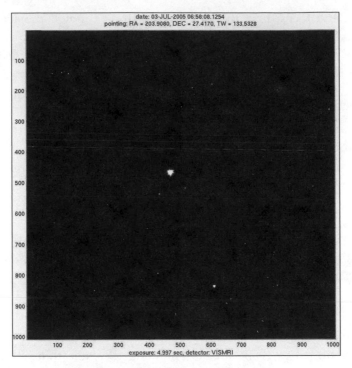

The impactor, at center, bids farewell.

Chapter 25

Encounter

Some stayed at their consoles all night; others went home or to suite hotels, sleeping fitfully. They trickled back in and watched numbers on screens and chatted and nodded and then had lunch. As the comet closed in, the spacecraft seemed to be enjoying a lazy Sunday in space.

On the afternoon of NASA's "Spectacular Day of the Comet," scientists, engineers and navigators met in a conference room with Monte Henderson, Rick Grammier, Keyur Patel and other mission leaders. The attitude control system was working beautifully, they agreed.

Contingency plans years in the making would go unused. The impactor was designed to meet its mark even if released 30 kilometers off target. The small spacecraft was within one kilometer of the team's best estimate of true aim. The engineers weren't giddy, but they were quietly amazed at how extraordinarily well things seemed to be going. All that was left was to choose which of three possible encounter scripts to run.

Even the best-case scenario was complicated by the brightness of the comet. Exposure time shortens with brightness; images from a sequence expecting a "dim" comet and finding a bright one would come back bleached. The team had prepared and tested different imaging schemes for dim, bright and average comets, with "average" being about 4 percent reflective, the darker-than-charcoal hue of Halley, Borrelly and Wild 2.

They went with "average." Then the last of many thousand Ball-JPL Deep Impact meetings adjourned.

Some hours later, Nick Mastrodemos, the JPL navigator, stood outside with Charlie Schira, the Ball attitude control engineer. Security badges and laminated yellow Deep Impact phone-list cards hung from their belts. Well-dressed people with VIP tags streamed in. Mastrodemos stood silently. Despite his Greek skin tone, he looked pale. Part of it was fatigue. Mastrodemos had been on the night shift and slept two hours before coming back in, walking the half-mile from home back to JPL because, in the past two days, he had forgotten his car and then his bicycle at JPL. Schira asked him what was wrong.

"The attitude determination is working so well that, now if anything goes wrong, it's going to be my fault," Mastrodemos answered.

It was true. His algorithms—mathematical routines—were the real brains behind AutoNav, the autopilot software, and it was his neck on the line should the spacecraft miscalculate.

"Nick, you've run ten-thousand simulations," Schira reassured.

"It's not enough."

It had to be enough, though. The separate spacecraft were ripping through space at 49,000 miles per hour, with the impactor a few hours from the comet. Deep Impact listened to its minders on the ground, but only to particular messages—among them strings of digits sent from JPL navigators, numbers expressing more and more accurate estimates of the spacecraft's and the comet's relative positions and orbital intentions. Then, two hours before impact, the autonomous navigation software Mastrodemos worked so hard to design and refine would cut even that channel, a benevolent hijacker taking control of two spacecraft until there was only one left.

The sun set. Reporters hunched over laptop computers in their conference room across from the von Karman auditorium. It was a holding pen for several dozen members of the media, a safe distance from the buildings in which Deep Impact's minders worked. With impact scheduled for 10:52 p.m., an hour more familiar to sportswriters than science journalists, those on deadline followed the sports reporter's playbook, writing the whole story, except for the ending, in advance. Documentary crews filmed over their shoulders. VIPs gathered in a glass-walled room overlooking the JPL Mission Operations Center, from which NASA dignitaries, politicians and even former *Lost in Space* TV mom June Lockhart would follow the night's events.

The world tuned in to Deep Impact. A giant screen on Waikiki Beach telecasted the NASA TV feed as night fell. Hawaiians gathered, first a few, then hundreds, then a crowd of 10,000, to watch the events unfold. A thousand more came to the Denver Museum of Nature & Science, where Lorna Hess-Frey, the Ball engineer central to Deep Impact's redesign four years earlier, spoke onstage about the mission. She even signed a few autographs, which made her chuckle. Another 1,500 crammed into the University of Colorado's planetarium and adjacent Sommers-Bausch Observatory, across campus from the Hale Sciences building in whose basement the saga leading to this hour began 57 years earlier. Similar gatherings happened at planetariums and observatories across the United States. And, one by one, millions logged on to the mission Web site.

Jeremy Stober on the big screen at JPL

Keyur Patel, left, Dan Kubitschek, center, and Rick Grammier, standing, watch and wait in Deep Impact mission control.

Hundreds of amateur astronomers aimed their telescopes north of Spica in the constellation Virgo; professionals at nearly every major observatory on Earth did the same. Hubble and Spitzer and Chandra and Swift; the Submillimeter Wave Astronomy Satellite; the Solar and Heliospheric Observatory; and Odin, a Swedish radio astronomy satellite all turned to the same spot. Rosetta, the European spacecraft, also watched, a welcome diversion on the 10-year trip to its own comet rendezvous.

The Deep Impact team was divided into three places at JPL. Two were adjacent in Building 230's Mission Support Area. About 30 engineers sat in three rows, many with their own notebook computers on the desk next to their flat-screen monitors. Behind them were desks for Grammier, JPL director Charles Elachi, and NASA managers. Directly behind Grammier hung an American flag and a giant Deep Impact mission logo. Everyone paged through a thick binder, a mission operations log Jennifer Rocca and others had prepared. It contained a minute-by-minute account of what should happen in everything from a best-case "green-light flow" scenario to contingency nightmares.

Someone produced a container of Planter's roasted peanuts, good-luck legumes in JPL mission control since 1964, when an engineer brought them in for the seventh, and first successful, Ranger moon-impact mission. Projectors painted images from the flyby and impactor cameras onto screens descended from the ceiling. A bright blob in the center grew with each passing hour.

Amy Walsh monitored spacecraft computers; Tom Golden watched software performance; John Valdez checked temperatures; Henderson, with no direct responsibilities this night, sat at the front of the room with instrument engineer Don Hampton and Jon Mah, whom Hampton had recruited into mission operations. They checked the status of instruments and the images sent home. Tom Bank, watching over the spacecraft systems in general, was still next to Jeremy Stober, the impactor-system lead. Against his notebook screen Stober had leaned a small snapshot of his young sons, both born during the long years he had worked toward this moment.

At some point in the evening, everyone in the room was wearing either a red or a blue golf shirts. Blue was the color for mission managers and those watching the flyby spacecraft; Stober and the others monitoring the impactor wore red. Their work would end at impact, if there were an impact. The red shirts were either an overt or subconscious nod to the *Star Trek* "Red Shirt phenomenon," in which red-shirted Starship Enterprise crewmen who beamed down anywhere near Captain Kirk tended to die violently.

JPL Mission operations people, who, mission after mission, dealt with vital if generic tasks such as Deep Space Network communications and data handling, worked in a larger, adjacent space. The teams saw each other through picture windows. Deep Impact engineers who had worked the day shift gathered in the larger room to watch.

The science team was in another building, having taken up temporary residence in the Mars Exploration Rover science team area. Only two scientists were missing. Mike A'Hearn, the principal investigator, was with the mission operations people. As principal investigator, he would confirm a hit or a miss. The second absent scientist was astronomer Karen Meech. She was halfway across the Pacific Ocean, high atop Hawaii's Mauna Kea. From there, she orchestrated an extended videoconference with observatories in Chile, Spain, Australia, Taiwan and elsewhere.

Two hours before impact, AutoNav took over both spacecraft. The hardware and software of Deep Impact controlled its own fate.

Now, with a fraction of its 268-million-mile journey left, the big telescope for the first time saw the body of the comet rather than just a hazy coma. From 46,000 miles away, Tempel 1 spanned a single pixel, then two, then four, then, 90 minutes before their paths would cross, the spacecraft recognized the area equivalent of Washington, D.C. as ten adjacent electronic dots of varying shades.

The impactor's larger thrusters were set to fire in the first of three trajectory-change maneuvers. The first would be the longest. At his console, Steve

Sodja, the Ball engineer behind the propulsion system, watched the numbers as they refreshed on his screen. The impactor's smaller thrusters had been emitting their butterfly burps since shortly after release, changing the spacecraft's orientation in the absence of reaction wheels. Now the big thrusters would fire for the first time, the impactor's aim reliant on their output. The impactor had been hot and cold and waiting in the harsh environment of space for the better part of six months.

At the 90-minute mark, deep inside the impactor's lone computer, AutoNav mulled the ten-dot white-grey pixilated blob of comet. AutoNav churned through its equations and, fractions of a second later, told the attitude control system where the spacecraft should move. Mike Hughes, Charlie Schira and Lew Kendall watched as their software commanded the thruster valves to open. Hydrazine flooded catalyst beds warmed by John Valdez's heaters. The chemical bubbled, smoked, then chain-reacted. The pulse lasted a few seconds and had, the accelerometers reported, done exactly what AutoNav had asked.

Yet the spacecraft was off course.

It had been off target before the maneuver, too. From the moment it was released, the impactor spacecraft had been headed to within a kilometer of the imaginary corridor through which navigators believed the comet would swoop at 10:52 p.m. California time on July 3. But no one had known where in the glob of cometary dust the comet itself would be. The impactor spacecraft had, in fact, been heading wide left the whole time, on a path to miss Tempel 1 by about a kilometer. Achieving such accuracy had been a feat, but AutoNav and the attitude control system would have to finish the job.

Mastrodemos and Dan Kubitschek watched their terminals with angst. Through headsets, colleagues reported status updates here and there, but everyone knew the mission now came down to the deep-space autopilot software. And something was amiss.

Mastrodemos stared at the numbers on his screen. The burn had been too big, and it had been in the wrong direction. AutoNav had corrected for the kilometer, alright, but rather than steering the impactor back toward the comet, it had pushed the spacecraft below Tempel 1 by about 3 kilometers.

Everyone on console had seen it before. It was known as a "zig."

In testing, the "zig" had been followed by a corrective "zag."

The zig-zagging had worried Mastrodemos and Kubitschek. It was the product of AutoNav confusing the slow rotation of the comet with the spacecraft itself slipping laterally. Rather than watching a turning carousel, AutoNav assumed it was walking around an idle one. So it told attitude control to tell propulsion to tell the impactor's big thrusters to correct a non-problem.

Mastrodemos and Kubitschek had proposed fixing the comet-rotation problem and even developed the software to do it. But the mission's leaders, now wearing blue shirts in this very room, had decided to leave it on the ground. And as long as they could count on a zag, the impactor was welcome to zig its heart out.

Mastrodemos had no philosophical issue with zigs or zags. He was, however, skeptical that zags inevitably followed zigs. The eleventh and final Deep Impact operational readiness test—in which the entire team had, in June, sat in these very chairs—ended with a zig, a zag and a righteous simulated *boom*. But earlier such tests had failed. They might have just gotten lucky in June.

A zag would have to come 35 minutes before encounter, with the impactor's second course correction. The comet grew in the big telescope's vision, adding a pixel to its snowy blob every couple of minutes. All the while, AutoNav's keepers watched in quiet horror as their software calculated and recalculated trajectories in its quest to conjoin the impactor's and comet's orbits for that one brief instant. For five minutes, ten minutes, fifteen minutes, AutoNav continued to insist not on a zag, but rather another zig, which could put the impactor so far away that the final adjustment, twelve and a half minutes before encounter, could never make up for it.

Mastrodemos was, as he later put it, "scared out of my pants." He didn't dare look back at Kubitschek, a row behind him, for fear of tipping off the rest of the team that something was wrong. He wouldn't have seen Kubitschek anyway. AutoNav's behavior was agitating him to the point he had left his Impactor Flight Director seat to splash water on his face.

But as abruptly as it had gone astray, AutoNav corrected itself. The comet grew large enough that rotation mattered less than sheer presence, and the software converged on a trajectory more in line with its true relative position in space. The second thruster pulse brought a gorgeous zag, pushing the impactor right at the heart of the comet. Then the flyby sent the impactor one last message, telling its partner precisely when the comet would mow it down. The little spacecraft changed the timing of its own imaging sequences such that its 16-frame-per-second crescendo would happen just as the comet approached, and not two minutes earlier or, if AutoNav kept her on course, never.

The big screens showed something with no resemblance to a banana or bowling pin or dumbbell or any of the other worst-case comets orbiting Mastrodemos's nightmares. This was a soaring avocado, a space spud, something much closer to a dinner roll than a donut. Scientists' conviction that Tempel 1 would be long and lean had been mistaken. Kendall and Schira continued to watch attitude control numbers roll in, but the comet itself became a serious distraction.

"Look at that," Schira said quietly. "We can hit that."

The comet grew faster, the pixels piling on and adding tantalizing detail. It had fewer divots than the Stardust mission's comet Wild 2, and what looked like smooth patches and scarps. Tom Golden, monitoring spacecraft software throughput at a desk directly under the big screens, couldn't help but look up at them. He, like others in the room, had been so focused on the spacecraft that they had put little thought into the object of the mission's desire. Yet there it was—their comet. Way out there in space. *Just look at that thing.*

Others from the Deep Impact team watched from points far from Pasadena. Michelle Goldman, the test engineer, looked to the skies from an observatory in Israel where she was vacationing. Joe Galamback and Alec Baldwin, the technicians who had put the spacecraft together, followed via NASA TV from their homes in metropolitan Denver and the foothills above Boulder. Tom Yarnell, with his family at the event at the University of Colorado observatory, considered the same feed with disbelief. It seemed bizarre to him that the telescopes he had helped build were 83 million miles away, taking these amazing pictures.

Out there, the two spacecraft had separated by more than the distance between New York and Honolulu. The flyby was snapping images and then subtly turning to scan the slit of its infrared detector across the body of the comet. The impactor took pictures for scientists, too, but more importantly, it plotted its final course correction. AutoNav's scene-analysis algorithm combined the brightness of spots on the comet with its understanding of where the flyby spacecraft was flying. With less than 13 minutes until its appointment with Tempel 1, still 4,800 miles from its target, the impactor puffed itself a bit south, toward a sunbathed area on the soft underbelly of the frozen world. The flyby, running the same AutoNav software, shifted its aim accordingly despite having no way of knowing what spot the impactor had chosen.

The subsystem engineers uttered the occasional update, to be heard over the headsets of everyone in the room and the NASA TV audience. But it was mostly quiet now, high anticipation damping the typical background rustling of three dozen people in a room. Pool photographers and a TV cameraman loomed over engineers' shoulders and knelt at their feet, seeking action shots but finding the only things moving to be numbers on screens.

Meanwhile, in Hawaii, Karen Meech had put on her wool hat and sweater and stepped into the evening chill on Mauna Kea, where she stood among amateur astronomers. The organizer of the greatest astronomical campaign in history had been instructed by her teammates in Pasadena to look at the sky without so much as binoculars. Should she perceive a visible flash, the idea was, she could tell the rest of them about it via the satellite phone on her hip. Given the distances involved, Meech knew she would see nothing more than starry sky, the impact too distant to reach her eyes.

But out among those stars, the drama unfolded, an imminent collision between a 20-billion-ton hunk of nature and an 820-pound creation of man. The little spacecraft shot image after image of the instrument of its doom, racing backwards as if trying to escape, but at a speed calculated to be way too slow to outrun Tempel 1. A looping journey ranging three times Earth's distance to the sun had shrunk to thousands and now just hundreds of miles. The comet was exploding into the impactor's view.

The small craft felt the sandblast of cometary dust against its carapace, then the hammer blows of particles moving 100 times faster than a rifle shot. No heavier than coarse grains of sand, they knocked the impactor off kilter; the attitude control system leapt into action to aim the camera back at its target zone on the nucleus. With just over three seconds until the comet would pass, the impactor snapped a black and white photograph from 23 miles above the comet. It depicted a crater and its environs, each pixel spanning about three feet. The impactor packaged it up into all of four kilobytes—the size of a short e-mail—and sent the most detailed image ever taken of a comet back to its partner.

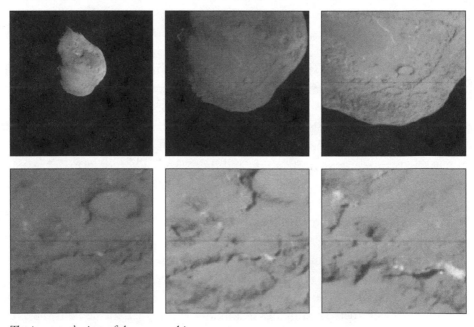

The impactor's view of the approaching comet

The impactor's fate was the flyby spacecraft's secret for the next seven and a half minutes. Whether in the form of light waves striking telescope lenses or the radio signals tickling the Deep Space Network's ears, the news took that long to arrive back home.

In mission control, Larry Murphy, Ball Aerospace's most seasoned telecom engineer, stared at his screen. At 62, he was the oldest person on Ball's Deep Impact team, a 33-year veteran of the company. He was retiring tomorrow, having decided to spend the last hours of his career wearing a red shirt in Pasadena. At 10:52 p.m., a couple of numbers went dead. Murphy blinked. They did not resurrect. He gulped in the room's silence.

Stober, the impactor systems engineer, was watching the same numbers on his own screen. He sat behind Murphy. He knew Murphy was supposed to say something like, "Communications with the impactor have been lost." Then the team of engineers was to play dumb and continue to monitor events as if the comet-hunter had not met its mark until given the official O.K. to celebrate.

The reticence was justified, Stober knew. A comet chunk could have tagged the spacecraft, sending it into a silent tumble just a shade too far south. Then you have the embarrassment of premature elation salting the agony of defeat.

Murphy was frozen in place, staring at the screen. *Larry. Larry. Laaaaary,* Stober was thinking. The older man wanted to leap out of his seat. Then he gathered himself and adjusted his headset. After an interminable few seconds, Murphy said, "Flight, this is telecom. We have lost lock with the impactor."

A minute passed, then another, and another. Engineers stared at their screens. The impactor's pulse had flatlined. The flyby's telescopes clicked away, the spacecraft's silicon synapses firing, its voice clear across all those millions of miles. But what was it seeing? Striking the comet was an engineering feat, but NASA hadn't spent $330 million on marksmanship. It had paid for science data, specifically a minimum of one image of the cosmic explosion. Mike A'Hearn's job was to find that image.

In the adjacent mission-support area, A'Hearn stared as shot after shot popped onto his screen. Was that a flash? No. Next. That had to be a flash. But was it an image artifact? He needed a bit more. Then came the image. At a distance of more than 5,000 miles from the impact, the big telescope had snapped a slightly blurry photo, as all its photos were. From Tempel 1's nether regions shot out a giant, bright cone nearly as large as the comet itself.

The impactor had indeed melted and atomized as it plunged into the comet body, which hurled it back into space in a strobe fountain of heat and light. Dust and frozen water and gases followed in prodigious volumes, after 4.6 billion years free at last.

Impact with the comet Tempel 1

The team of red- and blue-shirted engineers saw the same image at the same time A'Hearn did. Through the picture windows in the front of the room, on screens high above the cavernous Mission Support Area in which A'Hearn sat, was live video of Deep Impact scientists two buildings away. They were hugging and high-fiving in front of that still shot of a wounded comet. Mike Belton jumped up and down in front of the screens. Jessica Sunshine lifted her old Brown University advisor Pete Shultz right off the ground with a hug; Schultz, looking at the screen, was thinking: *just like the garden perlite!*

Amy Walsh watched it all through the windows and smiled. There was the impact image, an astonishing sight. But what really struck her were all those jubilant PhDs. That was it, she thought. We did it. We did our jobs. Now it's up to them to go analyze the data and figure out *what* we did.

"Yes, we've hit it," A'Hearn told Grammier through his headset; Grammier relayed the information to the team and the global NASA TV audience. But his words, "Team, we've got confirmation," were drowned in the cheering. Even engineers can suppress only so much.

Some, like Mike Hughes, jumped from their seats, fists raised in triumph. Stober hugged Bank. Grammier hugged anybody in the vicinity. They high-fived, patted backs, clapped, whooped. Mastrodemos slumped back in his chair, the weight on his shoulders finally lifted. He mustered a handshake with his neighbor.

Outside the glass, with the mission controllers in the larger room, the JPL mission test engineer Jennifer Rocca stared at the big screens with tears streaming down her face. She felt like the Super Bowl MVP, and the smartest person on Earth. She knew this was why she did this work. Why she made the sacrifice. She knew it would be fleeting—in this business, you spent your life proving to yourself how *not* smart you were, after all. She wondered how many people had a feeling like this in their entire lives.

At the fringes of the same room, David Taylor, the Ball Aerospace CEO, beamed. He couldn't have asked for a higher-profile success. Ball Aerospace, descended from a jar maker and a university research project, was now a player beyond the orbit confining it for more than 40 years. Then he noticed John Marriott, the manager whose toughness had made this success possible, hunched over next to him.

The man was crying uncontrollably—the enormous pent-up stress bursting free like so much cometary material. Marriott himself couldn't believe it. Well-wishers approached; his wife Beth and his boss ran interference. "Give him a few minutes," they begged. The business of building spacecraft had become too much for him, Marriott realized at that moment.

All the while, Larry Murphy stared at his screen, watching numbers. The surviving spacecraft was still out there, living up to its name, *flying by* the comet. It was Living for the Moment, slewing and shooting and storing, aiming and re-aiming its camera and antenna gimbal, and chucking images and spectra back at the bright dot of Earth as fast as it could, its computers saturated with activity, its data pipes bursting.

Grammier reminded everyone they still had a spacecraft to fly. Even those in red, the mission's newly jobless, sat back down at their terminals, and it was as if they had never gotten up. Keyboard taps and brief acronym-encoded status updates interrupted silence.

Thirteen minutes after the blinding flash, Deep Impact's sole remaining spacecraft took its last close-up of the comet from 435 miles away, the big telescope aiming for a look at the divot its partner had left. Then it turned its armored flank to the comet's wash before passing within 310 miles of the enormous black body.

Emerging unscathed, Deep Impact's survivor turned back to capture Tempel 1 as it shrank into the night, the comet's outlines silhouetted by a brilliant cloud of gas and dust radiating from a crater that humans had made.

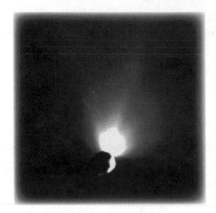

Epilogue

"**W**e touched a comet," the scientist Pete Schultz said in the early hours of July 4, 2005. "And we touched it hard."

Deep Impact was the world's top story later that day. The image of the impactor's strike became an overnight icon. Ed Weiler's cosmic explosion had captured the public's attention, just as he had predicted it would. The next day, The Comets, the band once led by rock & roll pioneer Bill Haley, played JPL's mission party.

Together with the ongoing successes of the twin Mars Exploration Rovers and the Stardust mission, which captured dust from the comet Wild 2 and deposited it back on Earth in January 2006, Deep Impact was part of a surge of good news that helped pull NASA out of its slump.

Magazines such as *Aviation Week & Space Technology* and *Sky & Telescope* put images of the wounded comet on their covers. The *New York Times* editorialized on the event, writing that "NASA scored a remarkable triumph this week when it smashed a small spacecraft into an onrushing comet with almost bull's-eye precision."[1] A NASA-record 8 million Web visitors had stopped by the mission Web site in the 24 hours surrounding impact.[2]

Dozens of scientific papers emerged as scientists analyzed Deep Impact's data, the first batch appearing three months later in a special edition of *Science*, the preeminent American scientific journal. Deep Impact's view of part of itself atomizing against the belly of Tempel 1 again landed on the cover.

The image belongs with a handful of photos depicting just how far earthly evolution has come in the 4.6 billion years since the ices and organic chemicals of comets began agglomerating into our rocky planet. In such an exclusive album belong the Apollo 8 "Earthrise"; the Voyager "pale blue dot" image of Earth from beyond the orbit of Pluto; Cassini's shot of Saturn aglow, backlit by the sun; and any one of a slew of images the Mars Exploration Rovers have snapped of their own caterpillar tracks on Martian soil. To capture such beauty took some of our most advanced technologies, some of our brightest people,

exquisite coordination, enormous riches, and, perhaps most importantly, a collective desire to explore. These photos are our pyramids.

As scientists had hoped, Deep Impact's imagery and spectroscopy led to a leap in our knowledge of comets. The impact dislodged huge amounts of microscopic dust—silicates similar to those making up the Earth's crust—showing Tempel 1 to be more a snowy dirtball than the dirty snowball. The magnitude of the comet's response to the small insult of Deep Impact led scientists to conclude that the body was extremely porous, the powdery dust at least tens of meters deep. Scientists wondered if the diverse features of the comet's surface (smooth here, pockmarked there) meant that two smaller comets had mashed together to make Tempel 1. The comet had certainly formed at very low speed, lending credence to prevailing theories that such bodies indeed jelled gently in the outer reaches of the solar system. Mike Belton came to believe that Tempel 1—and possibly other comets—may have formed in layers of "talps" ("splat" spelled backwards) piling on a hidden core rather than grain by grain.

The lookback images had been more than just beautiful. Scientists used the shining plume's rate of expansion to estimate the comet's mass and density. They weighed Tempel 1 in at about 20 billion tons and pegged the comet to be just 40 percent as dense as water, a fluffy snowbank (Earth, by comparison, is 5.5 times denser than water).[3] They figured the comet's gravity to be such that a grown man on its surface would weigh about much as a penny does here.

Jessica Sunshine, the Deep Impact chief spectroscopist, spotted water ice on the comet's surface, a first. Carey Lisse, the Applied Physics Laboratory astronomer on the Deep Impact science team, found evidence of carbonates, which must form in liquid water at moderate temperatures. Thus Tempel 1 somehow harbors ices frozen in the solar system's extremities, carbonates formed at moderate temperatures, and silicates formed at more than 1,000 degrees, the last of these also a key finding from the Stardust comet mission. It sent theorists of solar system formation back to their models.

Deep Impact failed to deliver on one of its main goals. It never captured an image of the crater. The sooty dust was so dense, reflective and persistent that it obscured the flyby's view. Schultz thought the crater was big—probably the diameter of two football fields—and that it might have been shaped like a sombrero. But its barbecuing by the sun, which scientists estimated cost the comet a billion tons of mass in 2005, probably reshaped it in short order, Schultz said.

The scientific work continued. The many academic papers emerging in the two years following were just the beginning, Mike Belton believed. A second round of science, combining information from the spacecraft's instruments and the Hubble Space Telescope, for example, could change the way we think about comets. For one thing, Belton said, Tempel 1 appears to have been physically and chemically "alive," internally active, with frozen volcanoes pouring pulverized ices from the depths.

With experiments come new theories beckoning more experiments. So goes the march of science. "In the picture that's been drawn from Deep Impact, the possibilities for new insights are just enormous. I don't know what they all are," Belton said.

The Ball Aerospace telecommunications guru Larry Murphy indeed retired, as did John Marriott, who in 2006, despite being only in his early fifties, opted for a life on an apple farm in upstate New York, not far from where he grew up. Monte Henderson, Marriott's former deputy on Deep Impact, took over another major Discovery program at Ball: Kepler.

Kepler emerged from a far-out idea NASA scientist Bill Borucki brought to Alan Delamere and Ball Aerospace in the early 1990s. Borucki dreamed of a space telescope looking for distant planets. But rather than film or an electronic sensor to capture images, the telescope would carry a metal plate with pinpoint holes to let through the light of sun-like stars. It was, Delamere recognized, unbuildable. But a bit of quick testing in Boulder showed that a CCD—the "film" of a digital camera—could be made to do the same thing. NASA rejected Kepler not once, but four times before finally being seduced in December 2001.

Ball set about building the spacecraft and its instrument; the University of Colorado's Laboratory for Atmospheric and Space Physics—the much-evolved Rocket Project, with its own modern complex not far from Ball Aerospace in east Boulder—would be in charge of mission operations. LASP had become become a space research powerhouse in its own right, with 65 faculty, 120 student research assistants, and more than 400 staff having built instruments for orbiters and spacecraft destined for seven of the eight planets. For the eighth, Mars, LASP is leading NASA's $485 million Mars Atmosphere and Volatile Evolution mission (MAVEN), scheduled for launch in 2013.

Across the CU campus from LASP, a towering glass case in the Hale Sciences building's carpeted basement hallway displays Cortez, Piedra and Mesa Verde pottery dating back 1,000 years. Down the hall hang scientific posters with such titles as "Commonalisms Between Humans and Chimpanzees in Southeastern Senegal." Office doors of anthropology professors and primate biologists, the modern lords of Hale, interrupt the chest-high wood paneling. Nowhere is even a hint of the role these former Rocket Project quarters played in shaping the history of Boulder and, more broadly, the American space program.

On the first Friday of March 2009, LASP hosted an open house for Kepler's night launch from the same Cape Canaveral pad from which Deep Impact had departed more than four years earlier. Not 30 feet from an Aerobee rocket hanging in the lobby, behind a floor-to-ceiling plate-glass window, sat Jeremy Stober. He wore a tie and a navy-blue sweater rather than a red polo shirt. On his desk was a thick mission operations binder, two flat-screen monitors presenting rows of color-coded data, and his notebook computer. Unlike the Deep Impact encounter, no photograph of his two young sons leaned against its screen. Stober, the Kepler mission operations manager, could simply turn around and see his boys, both a head taller now, watching him raptly among the crowd amassed on the other side of the glass.

Now settled into its Earth-trailing orbit, the 15-foot-tall space telescope's 95-megapixel digital camera is taking in light from the Cygnus-Lyra region near the plane of the Milky Way. Kepler's mission is to track about 100,000 sun-like stars for three-and-a-half years, watching for subtle hints of planets in habitable-zone "Goldilocks" orbits in which H_2O can exist as a liquid, a fundamental precondition for life. If Deep Impact was a space slugfest, Kepler is a staring contest with the stars.

The flying photometer was to cost $286 million and launch in 2006; by the time Kepler departed on March 6, 2009, for reasons spanning technical surprises to rocket availability, its tab had more than doubled. The lessons of the Rocket Project, the first Orbiting Solar Observatory, Deep Impact and so many other missions held true: doing something new in space is almost always much harder—and pricier—than anyone thinks it will be. However mature the space industry becomes, however sophisticated the management techniques, however intense the oversight, however brilliant the engineers, those breaking new ground in space engineering remain, as Nancy Grace Roman put it, like babies learning to walk. They will fall down.

Regardless, Deep Impact might be considered a bargain, according to the NASA manager who watched it most closely. "It was a pretty incredible project for that amount of money. It really was. Even if the final figure was $333 million," Lindley Johnson said. "To do that mission for that kind of money, it's commendable."

Monte Henderson spent two years leading the Kepler program for Ball, warding off cancellation despite all the delays and blown budgets. But in 2007, he was relieved of his duties, just as Brian Muirhead had been on Deep Impact. It was no easier for him than it had been for Muirhead. Henderson moved on to manage projects in Ball Aerospace's antenna business.

Henderson's JPL counterpart Rick Grammier became director of solar system exploration at JPL. Grammier's Deep Impact deputy Keyur Patel was tapped to be project manager of the Dawn Mission, which launched in September 2007. It is scheduled to visit two asteroids, Vesta in 2011 and Ceres in 2015.

Deep Impact, Mercury Messenger, Kepler and Dawn cost so much that NASA's Discovery Program could afford no new missions for six years. In late 2007, NASA finally selected the Gravity Recovery and Interior Laboratory (GRAIL) moon orbiters. Ball had backed several of the Discovery Program proposal teams that year. For the next round of Discovery proposals, scheduled for late 2010, Ball is supporting no fewer than 10 missions.

"We're still hoping to be able to capitalize on being able to do Discovery missions," said Carol Lane, Ball Aerospace's vice president of Washington operations. "It's right in our sweet spot."

For Ball Aerospace, the business boost expected in the wake of Deep Impact never materialized, a reflection of broader economic trends and lost proposal bids. NASA's budget has been flat since Deep Impact, and the space agency's science endeavors by 2010 had $1 billion less to work with than the $5.5 billion they enjoyed in 2005. Rather than expand their business into deep space, Ball Aerospace was forced into repeated rounds of layoffs. The company was doing more military work, which amounted to 55 percent of its business in 2009. The cultivation of relationships with space scientists suffered in this environment, and budget pressures tempered customer tolerance for creative—

Willie Bottema, with the COSTAR instrument her husband Murk designed, at the Smithsonian National Air & Space Museum in November 2009

and often expensive—engineering for which the company has been known since the 1950s. How much of the "old Ball" survives as part of the "new Ball" remains to be seen.

Ball built the final two instruments for the Hubble Space Telescope, the Wide Field Camera 3 and the University of Colorado-designed Cosmic Origins Spectrograph, installed during the final Hubble servicing mission in May 2009. The spectrograph took residence in the bay Murk Bottema's COSTAR eyeglasses had occupied since 1993. COSTAR had been idle for years, its prescription built into replacement instruments. In November 2009, Bottema's widow, Willie, attended the COSTAR exhibit's dedication ceremony at the Smithsonian National Air & Space Museum in Washington, D.C., as Ball Aerospace's guest of honor. All told, Ball built seven Hubble instruments, including all five aboard until sometime after 2014, when Hubble's storied career is scheduled to end.

The extra millions NASA spent for Deep Impact to turn the new RAD-750 microprocessor into a space-computing workhorse proved to be a good investment. More than 20 spacecraft, civilian and military, were later built around the faster processors.

Jennifer Rocca, the young JPL engineer, and Amy Walsh, who won Women in Aerospace's 2006 Outstanding Achievement Award for her work on Deep Impact, soon had young daughters. Both remained dedicated to their employers, but neither woman was willing to sacrifice to the extremes they did on Deep Impact. Another crop of the young and enthusiastic had to rise to the occasion. In this respect, space missions are like wars: the old start them; the young fight them.

And the Deep Impact spacecraft? It had been hibernating peacefully in an orbit it could have ridden for a billion years. But here was a fully tested spacecraft at the scientific world's disposal, and with half its fuel left. NASA opened bidding for what it called "missions of opportunity" involving the reuse of pre-owned spacecraft. Two particularly interesting prospects emerged.

One, called Extrasolar Planet Observations and Characterization, or EP-OCh, was proposed by Goddard Space Flight Center scientist Drake Deming. He wanted to aim the spacecraft's big telescope at a handful of stars to look for planets. The liability of the big telescope's blurriness was now an asset. Light slopping across its sensor would boost its sensitivity, which is critical to extra-solar planet hunts.

Mike A'Hearn proposed the second mission, called the Deep Impact eX-tended Investigation, or DIXI. The mission was to send the Deep Impact flyby to the comet Boethin in late 2008, adding a fifth comet to the four observed up close since 1986.

NASA liked both ideas and combined them into the $33 million EPOXI mission. Many from the Deep Impact science team moved to the new effort, including Alan Delamere, retired from Ball Aerospace but busy as a consul-tant. In 2008, the High Resolution Instrument spotted several planets that had revealed their presence through subtle dips in brightness evident when they transited across the face of their stars.

When the planet-watching ended, NASA enlisted the Deep Impact spacecraft for the first test of its "interplanetary Internet," which used pro-tocols evolved from the terrestrial network to transfer data among spacecraft and ground systems despite light-time delays that can span hours. The system worked, with dozens of images successfully transferred to and from the space-craft.[4]

Then comet Boethin failed to turn up. With NASA's blessing, A'Hearn changed its target to the comet Hartley 2. Deep Impact flew by Earth in late 2009, and was scheduled to zip past Hartley 2 on November 4, 2010. Once it sends its observations home, fourteen years after it was conceived, Deep Impact will be done.

"A scientific spacecraft is ephemeral. It goes away. But its results are mat-ters of knowledge. There are lasting effects we can see," said Dick Woolley, who worked on both the Orbiting Solar Observatory and Deep Impact. "My gosh, a guy can feel that he has done something in the world, anyway."

Woolley died of a lung ailment in August 2008, a few weeks after speak-ing these words. David Stacey, James Jackson, Russ Nidey and so many other Rocket Project and Ball Aerospace pioneers were already gone.

Nidey had lasted only a couple of years at Ball, joining the nascent Kitt Peak National Observatory southwest of Tucson, Arizona, in mid-1959. As

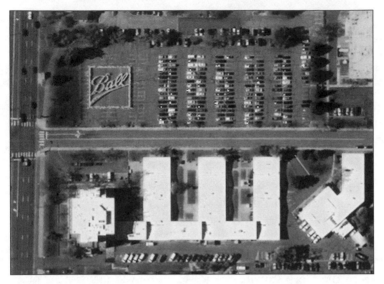

The Ball Aerospace "Living Logo." To celebrate the company's 50th anniversary, about 1,000 Ball employees wearing white shirts and ballcaps stood in a Ball parking lot just after noon on October 23, 2006, to wait for the DigitalGlobe QuickBird 2 satellite, which the company built. From 280 miles above and traveling at about 18,000 mph, the satellite took this shot. Arapahoe Road is to the left. The original Control Cells Corp. building, much expanded, is in the lower-right-hand corner of the photo, with the Tech Tower to the lower left.

colleagues in Boulder began working on the Orbiting Solar Observatory, Nidey was to lead the engineering of a space telescope orbiting "26,000 miles in space over Tucson sending messages about what it sees," as a newspaper account described it.[5] When that project proved overambitious, Nidey launched a sounding rocket program at Kitt Peak, then worked as systems manager of the observatory's Space Division, and, later, as a consultant. Before succumbing to Alzheimer's disease in 2008, he lived in a ramshackle 1970s-era mobile home in Cottonwood, Arizona, taught at Yavapai College, built large contraptions with astronomical intent, and generally enjoyed being called "professor." Late in his life, a friend asked him whether he regretted having lost his arm all those years ago. "Of course," Nidey answered. "It's hard to hold your wife in your arms when you have only one arm."[6]

Cornell astronomer Joe Veverka, the man who led the ill-fated Contour mission and went on to be a Deep Impact scientist, would have another shot at leading a comet hunt. Veverka would steer the surviving Stardust comet spacecraft toward Deep Impact's target. Originally called ScarQuest, the $24-million Stardust NeXT mission is scheduled to fly within 120 miles of Tempel

1 on February 14, 2011, during the comet's next visit to the sun. Finally getting a look at the impactor's crater would be a bonus, but the main goal is to observe changes to the comet's surface from solar heating, landslides, shifting of material, and impacts.

Many former Deep Impact engineers went on to work on WorldView-2, a commercial Earth-observing satellite; others moved to the Wide-field Infrared Survey Explorer, or WISE, a telescope cooled with hydrogen so frigid it's a solid. WISE began surveying the sky in 2009 in preparation for the James Webb Space Telescope, NASA's encore to Hubble. Ball is building the mirror array for James Webb, scheduled to launch in 2013. Comprised of 18 hexagonal segments, each more than four feet (1.3 meters) wide, the array is to fold into a rocket nose and open into a single mirror more than 21 feet (6.5 meters) across. Actuators will then adjust the mirrors in temperatures as cold as minus-400 degrees. The space telescope is designed to be 100 times more sensitive than Hubble in the infrared, promising scientists the deepest-ever views into the universe's past.

All this grew from Ed Ball's dream of diversifying his family's aging Midwestern glass company. His Boulder adventure had gone from analog machines to flying computers, studying the brightest object in the sky, dark matter, and everything between. Ed Ball witnessed many of his research division's successes before he died in 2000, at the age of 95. On the floor of the U.S. Senate, Indiana Senator Richard Lugar called him "one of our greatest citizens," adding, "His type of man will not be seen anytime soon, if ever." The *Muncie Star Press* called Ball "Muncie's man of the millennium."[7] He was certainly one of Boulder's, too.

Among those involved with the mission, a certain nostalgia set in. Never again in their careers would they be involved in something quite like Ball's deep-space debut. As Nick Taylor had promised Jon Mah, they indeed looked back on their time spent with Deep Impact as "the good old days."

NOTES

Chapter 1

[1] William B. Pietenpol, *Upper Air Laboratory, Progress Report Number 10, Contract No. W19-122ac-9, 31 October 1950* (Boulder: University of Colorado, 1950), 9.

[2] Milton W. Rosen, *The Viking Rocket Story* (Harper & Brothers, New York, 1955), 2.

[3] Charles P. Smith Jr., *Upper Atmosphere Research Report No. XXI: Summary of Upper Atmosphere Rocket Research Findings*, NRL Report 4276, February 1954 (Supplemented February 1958), 441.

[4] Ibid.

[5] Frank Kreith, personal interview, May 8, 2007.

[6] William E. Burrows, *This New Ocean: The Story of the Space Age* (New York: Random House, 1998), 102.

[7] Frederick I. Ordway III and Mitchell R. Sharpe, foreword by Wernher von Braun, *The Rocket Team* (New York: Thomas Y. Crowell, 1979), 274.

[8] David H. DeVorkin, *Science with a Vengeance: How the Military Created the US Space Sciences After World War II* (New York: Springer-Verlag, 1992), 30.

[9] Ernst Klee and Otto Merk, *The Birth of the Missile: The Secrets of Peenemünde* (New York: E.P. Dutton, 1965), 7.

[10] Tim Hunter, "The Astrophysics of Sunburns," The Grasslands Observatory, http://www.3towers.com/sGrasslands/Essays/Sunburn/Sunburn01.asp (accessed March 2, 2009).

[11] DeVorkin, *Science with a Vengeance*, 142-144.

[12] Ibid, 112.

[13] An alternate take on the Rocket Project creation story is that Walter Orr Roberts's Harvard College Observatory boss Donald Menzel recommended the University of Colorado physics department to Air Force Cambridge director Marcus O'Day, who then dispatched Miley. Menzel and Roberts were familiar with the CU physics department, which had been a partner with Harvard on the High Altitude Observatory since 1946. See David N. Spires, "Walter Orr Roberts and the Development of Boulder's Aerospace Community," *Quest: The History of Spaceflight Quarterly* Vol. 6, No. 4 (Winter 1998), 5-13. The Miley-Pietenpol scenario finds support in Albert A. Bartlett and Jack J. Kraushaar, *The Department of Physics of the University of Colorado at Boulder, 1876-1996* (Boulder: University of Colorado, 1996), 11-24.

[14] Richard Woolley, personal interview, July 3, 2007.

[15] DeVorkin, *Science with a Vengeance*, 75.

[16] Albert A. Bartlett and Jack J. Kraushaar, *The Department of Physics of the University of Colorado at Boulder, 1876-1996* (Boulder: University of Colorado, 1996), 14-1.

[17] Palmer Carlin, personal interview, February 17, 2007. Carlin, later a CU professor of electrical engineering, worked for Pietenpol while pursuing his physics PhD in the late 1940s and early 1950s.

[18] DeVorkin, *Science with a Vengeance*, 221.

[19] William B. Pietenpol, *Airborne Coronagraph for Rocket Installation, Progress Report No. 1, April 1, 1948 to July 1, 1948* (Boulder: University of Colorado, 1948), 7.

[20] Ibid, 8.

[21] National Center for Atmospheric Research High Altitude Observatory, "HAO History Photos," http://www.hao.ucar.edu/hao/history/pictures.html (accessed October 6, 2009).

[22] "Yngve Öhman," *Solar Physics* 119, no. 1 (1989), iv.

[23] Pietenpol, *Progress Report no. 1, 1*, 14.

[24] William B. Pietenpol, *University of Colorado Upper Air Laboratory, Progress Report No. 9, Contract No. W19-122ac-9, July 31, 1950* (Boulder: University of Colorado, 1950), 15.

Chapter 2

[1] Palmer Carlin, personal interview, February 17, 2007.

[2] William Simons, personal interview, May 4, 2007.

[3] David A. Mindell, *Between Human and Machine: Feedback, Control, and Computing* (The Johns Hopkins University Press: Baltimore, 2002), 179.

[4] Upper Air Laboratory, *Financial Report of the Upper Air Laboratory, University of Colorado, 1 August 1951* (Boulder: University of Colorado, 1951), 14.

[5] DeVorkin, *Science with a Vengeance*, 224.

[6] William B. Pietenpol, *University of Colorado Upper Air Laboratory, Progress Report No. 12, Contract No. W19-122ac-9, April 30, 1951* (Boulder: University of Colorado, 1951), 11.

[7] Ibid, 17.

[8] Upper Air Laboratory, *Financial Report*, 6.

[9] Ibid.

[10] Pietenpol, *Progress Report no. 12*, 13-15.

[11] Ibid, 18.

[12] Otto Edwin Bartoe, "Mechanical Design of a Rocket Borne Solar Pointing Control" (master's thesis, University of Colorado, 1954), 11.

[13] Upper Air Laboratory, *Financial Report*, 35; Days accrued actually were higher than those cited, given the financial report's having been distributed after the July 1951 shutdown mentioned in Pietenpol, *University of Colorado Upper Air Laboratory, Progress Report No. 13, Contract No. W19-122ac-9, July 31, 1951* (Boulder: University of Colorado, 1951), 2.

[14] Pietenpol, *Progress Report no. 13*, 13.

[15] Upper Air Laboratory, *Financial Report*, 4.

[16] Ibid, 9.

[17] Ibid, 3.

[18] William Rense, transcript of interview by David DeVorkin, July 27, 1983, *Space Astronomy Oral History Project*, National Air and Space Museum, Washington, D.C., 23.

[19] William B. Pietenpol, *University of Colorado Upper Air Laboratory, Biaxial Pointing Control Development Final Report, Contract No. W19-122ac-9, November 30, 1953* (Boulder: University of Colorado, 1953), 10.

[20] William A. Rense, *Upper Air Laboratory Final Report: Instrumentation Phase of Extreme Ultraviolet Research, Contract Number: Air Force W19-122 Ac-9, 31 May 1952* (Boulder: University of Colorado, 1952), 8.

[21] Terry Jean Wallace, telephone interview, May 6, 2010 (purple nail polish); Bartoe, "Mechanical Design of a Rocket Borne Solar Pointing Control," 69.

Chapter 3

[1] Sylvia Pettem, "Bureau of Standards celebrates 50 years in Boulder," *Daily Camera*, September 9, 2004, 1D.

[2] "Control Cells are Unique, Electronic Weighing Devices," *Boulder Daily Camera*, June 27, 1956, 13.

[3] *Denver Post*, "Tested at Boulder: Slender 5-Inch Gadget Can Weigh Bottle of Perfume or Brick House," May 5, 1956.

[4] Laurence T. Paddock, "Control Cells Corp. Is A New, But Growing Boulder Company With A Unique Product, Unique Organization," *Boulder Daily Camera*, April 29, 1955, 16.

[5] *U.S. News and World Report*, "Is 'New Era' Really Here?" May 20, 1955, 21. In Robert Schiller, *Irrational Exuberance*, 2nd Ed. (Princeton: Princeton University Press, 2005), 115.

[6] Earl L. Conn, *Beneficence: Stories about the Ball Families of Muncie* (Muncie: Minnetrista Cultural Foundation, Inc., 2003), 23.

[7] Frederic A. Birmingham, *Ball Corporation: The First Century* (Indianapolis: Curtis Publishing, 1980), 69.

[8] Robert S. Lynd and Merrell Lynd, *Middletown in Transition: A Study in Cultural Conflicts* (New York: Harcourt, Brace and Company, 1937), 74. Like its famous 1929 predecessor, *Middletown*, the book was about Muncie, a city the authors chose for its being "as representative as possible of contemporary American life" in geographic, demographic and cultural terms.

[9] Ibid., 75.

[10] "Beneficence," the last work of Daniel Chester French, sculptor of the Lincoln Memorial in Washington, D.C., still stands on the Ball State campus, where students call her "Benny."

[11] *Edmund F. Ball: Uncommon Man* (Muncie: Ball Corporation, 1988), 24.

[12] Edmund F. Ball, *Staff Officer with The Fifth Army: Sicily, Salerno and Anzio* (New York: Exposition Press, 1958), 5.

[13] *Ed Ball's Century*, video, directed by Samuel W. Clemmons and Nancy B. Carlson (Muncie: Ball State University, 2000). Accessible online at http://dvisweb1.bsu.edu/media/contentdm/ed_ball_320x240.asx.

[14] Birmingham, *Ball Corporation*, 142; John W. Fisher, telephone interview, December 2, 2008.

[15] Susan B. Carter et al Eds., Table Dd366-436, "Physical output of selected manufactured products: 1860–1997," *Historical Statistics of the United States, Millennial Edition On Line* (Cambridge: Cambridge University Press, 2006) (accessed online August 23, 2007).

[16] Edmund F. Ball, *Staff Officer*, 358; Richard Blodgett, *Signature of Excellence: Ball Corporation at 125* (Old Saybrook: Greenwich Publishing Group, 2005), 33; John W. Fisher, telephone interview, December 2, 2008.

[17] Alexander E. Bracken, "The Father of Ball Aerospace," *Ball Aerospace Explore*, Issue 4, 1999, 6; Conn, *Beneficence*, 315.

[18] Birmingham, *Ball Corporation: The First Century*, 144.

[19] Conn, *Beneficence*, 301.

[20] R.A. Gaiser, untitled memo to W.C. Schade, September 12, 1955 (Edmund F. Ball files, Minnetrista Heritage Collection, Muncie, Indiana).

[21] Ibid.

[22] Minutes of the Regular Quarterly Meeting of the Board of Directors, Ball Brothers Company Incorporated, October 25, 1955 (E.F. Ball Files, Minnetrista Heritage Collection).

[23] W.C. Schade, untitled memo to E.F. Ball, November 25, 1955 (Edmund F. Ball files, Minnetrista Heritage Collection).

[24] Edmund F. Ball, letter to E. S. Safford, December 15, 1955 (E.F. Ball Files, Minnetrista Heritage Collection).

[25] John W. Fisher, letter to W.C. Schade, June 14, 1956 (E.F. Ball files, Minnetrista Heritage Collection).

[26] Edmund F. Ball, letter to E. S. Safford, August 27, 1956 (E.F. Ball files, Minnetrista Heritage Collection).

Chapter 4

[1] Robert Arentz, a Ball Aerospace manager and the company's unofficial historian, confirmed the initial meeting of Ed Ball, David Stacey and Ed Safford at the Stacey home in interviews with Stacey and Safford in 1999.

[2] Stacey and colleague James Jackson earned $4,500 per year as of November 1, 1950; Albert Bartlett, who joined the faculty as an assistant professor in September 1950, earned $4,200. Albert A. Bartlett and Jack J. Kraushaar, *The Department of Physics of the University of Colorado at Boulder, 1876–1996* (Boulder: University of Colorado, 1996), 14-9.

[3] William Rense, transcript of interview by David DeVorkin, July 27, 1983, *Space Astronomy Oral History Project*, Smithsonian National Air and Space Museum, Washington, D.C., 33.

[4] "National Science Foundation History," National Science Foundation, http://www.nsf.gov/about/history/overview-50.jsp (accessed August 28, 2007).

[5] DeVorkin, *Science with a Vengeance*, 344.

[6] William Rense, personal interview, April 21, 2007; Rense, transcript of interview by David DeVorkin, 51.

[7] Edmund F. Ball, letter to W.C. Schade and R.A. Gaiser, September. 21, 1955 (E.F. Ball Files, Minnetrista Heritage Collection).

[8] R.A. Gaiser, letter to E.F. Ball, September 23, 1955 (E.F. Ball Files, Minnetrista Heritage Collection).

[9] Edmund F. Ball, "Introduction," in Birmingham, *Ball Corporation: The First Century*, 15.

[10] Edmund F. Ball, letter to E.S. Safford, July 5, 1956 (E.F. Ball Files/Control Cells 1956–1957, Minnetrista Heritage Collection).

[11] Ibid.

[12] Merle Reisbeck, personal interview, June 12, 2007 ("unimaginably generous"; "swayed decisions"); O.E. Bartoe, personal interviews, November 9, 2006, December 2, 2006; July 24, 2007.

[13] Ward Darley, letter to Edmund F. Ball, November 9, 1956, (E.F. Ball Files, Minnetrista Heritage Collection).

[14] E.F. Ball, letter to Ward Darley, November 19, 1956 (E.F. Ball Files, Minnetrista Heritage Collection).

[15] O.E. Bartoe, personal interview, November 9, 2006.

[16] "A Company's Research Outlay May Reveal a Lot About It," *Business Week*, January 28, 1956, 126.

[17] R. A. Gaiser, memo to J.M. Jackson, October 29, 1957 (R.A. Gaiser Files, Minnetrista Heritage Collection).

[18] Rick Shaffer, telephone interview, July 20, 2007.

[19] James Jackson, David Stacey and O.E. Bartoe, "A Brief Technical Survey of the Control Cells Systems," Ball Brothers Research Corporation, January 18, 1957, 2-3.

[20] Ibid., 15.

[21] E.S. Safford, letter to Edmund F. Ball, February 18, 1957 (E.F. Ball files, Minnetrista Heritage Collection).

[22] The know-how had arrived with several refugees from Mount Sopris Instruments, which had made such devices in Boulder before going bankrupt.

[23] R.A. Gaiser, "Ball Brothers Research Corporation: 1958 Summary, 1959 Forecast, Recommendations," January 19, 1959 (R.A. Gaiser Files, Minnetrista Heritage Collection), 22.

[24] R.A. Gaiser, memo to J.M. Jackson, October 29, 1957 (R.A. Gaiser files, Minnetrista Heritage Collection).

[25] Edmund F. Ball, "President's Report to Stockholders, Ball Brothers Company Incorporated Annual Meeting," March 19, 1959, (E.F. Ball files, Minnetrista Heritage Collection), 12.

[26] Edmund F. Ball, "President's Report to Stockholders, Ball Brothers Company Incorporated Annual Meeting," March 20, 1958, (E.F. Ball Files, Minnetrista Heritage Collection), 8.

[27] R.A. Gaiser, "Ball Brothers Research Corporation: 1958 Summary, 1959 Forecast, Recommendations," 21.

[28] Merle Reisbeck, personal interview, June 12, 2007.

[29] R. C. Mercure, personal interview, January 21, 2007.

Chapter 5

[1] Alfred Bester, *The Life and Death of a Satellite* (London: The Scientific Book Club, 1966), 31.

[2] Ibid, 41.

[3] Nancy Grace Roman, telephone interview, August 10, 2007.

[4] Newman Bumstead, "Rockets Explore the Air Above Us," *National Geographic*, April 1957, 566.

[5] Richard Woolley, "Why We Put Spectrographs in Orbit: A Justification of Science," speech to the Colorado-Wyoming Junior Academy of Science, December 6, 1969.

[6] Cargill R. Hall, "Origins of U.S. Space Policy: Eisenhower, Open Skies and Freedom of Space," in *Exploring the Unknown: Selected Documents in the History of the U.S. Civil Space Program, Vol. I: Organizing for Exploration*, (Washington D.C.: NASA History Office, 1995), 213-229.

[7] Ibid., 221.

[8] John E. Naugle, *First Among Equals: The Selection of NASA Space Science Experiments* (Washington, D.C.: NASA SP-4215, 1991). Available at http://history.nasa.gov/SP-4215/ch1-1.html#1.1.5 (accessed July 30, 2007).

[9] John E. Naugle and John M. Logsdon, "Space Science: Origins, Evolution and Organization," in John M. Logsdon, Ed., *Exploring the Unknown: Selected Documents in the History of the U.S. Civil Space Program, Vol. V: Exploring the Cosmos*, (Washington, D.C.: NASA History Office, 2001), 8.

[10] Ibid., 8; Nancy Grace Roman telephone interview, August 10, 2007.

[11] Karl Hufbauer, *Exploring the Sun: Solar Science Since Galileo* (Baltimore: The Johns Hopkins University Press, 1993), 168.

[12] Homer E. Newell, *Beyond the Atmosphere: Early Years of Space Science* (Washington, D.C.: NASA SP-4211, 1980), 246.

[13] John E. Naugle and John M. Logsdon, "Space Science: Origins, Evolution and Organization," in Logsdon, *Exploring the Universe*, Vol. V, 9.

[14] Hufbauer, *Exploring the Sun*, 168.

[15] Bester, *Life and Death of a Satellite*, 42.

[16] R.A. Gaiser, "Ball Brothers Research Corporation, 1958 Summary, 1959 Forecast Recommendations," January 19, 1959 (R.A. Gaiser files, Minnetrista Heritage Collection), 25.

[17] "Ball Brothers," transcript of interview of R.A. Gaiser, R.C. Mercure, O.E. Bartoe and Fred Dolder by David DeVorkin at the National Air and Space Museum, July 26, 1983, *Space Astronomy Oral History Project*, Smithsonian National Air and Space Museum, Washington, D.C., 27.

[18] Ibid., 27.

[19] Bartoe also predicted the need for what would be the world's first nutation damper, an oil-filled device designed to quell the slight wobbling one can observe as tiny loops amid the larger precessing circles drawn by a child's top. He calculated that nutation could otherwise pile upon itself and throw the satellite out of control.

[20] Based on the one-degree pointing accuracy of the state-of-the-art Discoverer/Corona spy satellites, which flew on the Agena booster.

Chapter 6

[1] E.F. Ball "President's Report to Stockholders, Ball Brothers Company Incorporated Annual Meeting, March 19, 1959" (E.F. Ball Files, Minnetrista Heritage Collection), 12.

[2] "Ball Brothers," SAOHP interview, July 26, 1983, 52.

[3] Ibid.

[4] NASA Discussion Group on Orbiting Solar Observatory Project, "Minutes of Meeting: NASA Headquarters, Washington, D.C., 23 May 1959," in John M. Logsdon, Ed., *Selected Documents in the History of the U.S. Civil Space Program, Volume VI: Space and Earth Science*, 58. (Washington, D.C.: NASA History Office, 2004), 58. Available online at http://history.nasa.gov/SP-4407/vol6/vol6.pdf.

[5] Frederick P. Dolder, "Solar Activity and the Yellow Coronal Line 5694 Å," (master's thesis, University of Colorado, 1952).

[6] Fred Dolder and Reuben H. Gablehouse, "Ball Bros. in Boulder, Colorado," transcript of interview by David DeVorkin, January 13, 1982, *Space Astronomy Oral History Project*, Smithsonian National Air and Space Museum, Washington, D.C., 11.

[7] Ibid.

[8] Pete Bartoe, personal interview, July 23, 2007; "Ball Bros. in Boulder," SAOHP interview, January 13, 1982, 5.

[9] "Ball Bros. in Boulder," SAOHP interview, January 13, 1982, 11-12.

[10] Bester, *Life and Death of a Satellite*, 68.

[11] NASA Facts (B-62): "Orbiting Solar Observatory: First of the 'Streetcar' Satellites," (Washington, D.C.: U.S. Government Printing Office, 1962), 1.

[12] R.C. Mercure, telephone interview, January 9, 2007.

[13] Wayne Biddle, *Barons of the Sky: From Early Flight to Strategic Warfare: The Story of the American Aerospace Industry* (New York: Simon & Schuster, 1991), 65.

[14] Ibid., 271.

[15] Joan Lisa Bromberg, *NASA and the Space Industry* (Baltimore: Johns Hopkins University Press, 1999), 19.

[16] Ibid., 30.

[17] RCA built the TIROS, the world's first weather satellite; Ford built the moon capsule for the Jet Propulsion Laboratory's Project Ranger; Bell Labs built Telstar, the world's first communication satellite.

[18] Bromberg, *NASA and the Space Industry*, 23-25.

[19] Woolley had served during the battles of Iwo Jima and Okinawa before his 21st birthday—on a transport ship going in that was a hospital ship coming out. More than 50 years removed, Woolley still choked up at the thought of it.

[20] Goddard Space Flight Center, *Orbiting Solar Observatory Satellite: OSO I The Project Summary*, NASA SP-57, (Washington, D.C.: National Aeronautics and Space Administration Scientific and Technical Information Division, 1965), 7.

[21] Ibid., 145.

[22] Ibid., 47-67.

[23] Bester, *Life and Death of a Satellite*, 74; new car prices from *The People History*, http://www.thepeoplehistory.com/60scars.html (accessed November 20, 2007).

[24] NASA SP-57, 193. The 950-pounder was a Bendix G-15; later came a CDC 1604, one of the first machines developed by Seymour Cray of later supercomputer fame. Cray's machine had 32 kilobytes of main memory, enough for about two seconds of song in a modern digital music player.

[25] Marion Fulk, telephone interview, July 28, 2007.

[26] Ibid.

[27] R.A. Gaiser, "Ball Brothers Research Corp.: Analysis-Evaluation—Forecast and Requirements," November 12, 1961, (R.A. Gaiser files, Minnetrista Heritage Collection), 5.

[28] E.F. Ball, *From Fruit Jars to Satellites: The Story of Ball Brothers Company Incorporated* (New York: The Newcomen Society of North America, 1960), 19.

[29] R.A. Gaiser, "Ball Brothers Research Corp.: Analysis-Evaluation—Forecast and Requirements," 71.

Chapter 7

[1] Lane E. Wallace, *Dreams, Hopes, Realities. NASA's Goddard Space Flight Center: The First Forty Years*, NASA SP-4312 (Washington, D.C.: NASA History Office, 1999), 60. Accessible online at http://history.nasa.gov/SP-4312/sp4312.htm.

[2] Barbara Dolder and Genevieve Gablehouse, personal interview, June 16, 2007.

3 "Ball Bros. in Boulder," SAOHP interview, January 13, 1982, 10.

4 L. T. Ostwald, personal interview, June 21, 2007.

5 R.C. Mercure, "Investigations of Solar Lyman Alpha Emission" (PhD dissertation, University of Colorado, 1957), 77.

6 R.C. Mercure, personal interview, January 17, 2007.

7 "Growth: 1960 Style," in *Ball Aerospace: The First Forty Years,* (Boulder: Ball Aerospace, 1996), 10.

8 Nancy Grace Roman, telephone interview, August 10, 2007.

9 Howard E. McCurdy, *Faster, Better, Cheaper: Low-Cost Innovation in the U.S. Space Program* (Baltimore: Johns Hopkins University Press, 2001), 81-82.

10 "Ball Bros. in Boulder," SAOHP interview, January 13, 1982, 14-15.

11 Reuben H. Gablehouse, remarks in VHS videotape of Ball Quarter Century Club Dinner in Boulder, 1988 (Genevieve Gablehouse collection).

12 "Coast guard" from NASA SP-57, 274; "solar panels" from NASA SP-57, 81; price per watt from Bester, *Life and Death of a Satellite*, 72.

13 Werner Neupert, personal interview, July 11, 2007; "Ball Bros. in Boulder," SAOHP interview, January 13, 1982, 9.

14 "Ball of Fire," *Business Week*, October 8, 1962 (reprint).

15 NASA, "News media conference, OSO I spacecraft performance," March 13, 1962 (NASA History Division, Orbiting Solar Observatory files).

16 Gene Lindberg, "Boulder's OSO 'Working Like Jewel,'" *Denver Post*, March 8, 1962, A27.

17 "To See The Sun," *Time Magazine,* March 16, 1962 (accessed online August 14, 2007); Martin Mann, "The march of SCIENCE," *Popular Science,* June 1962, 19.

18 Gene Lindberg, "Fruit Jar Firm Builds 'Moon,'" *Denver Post*, March 11, 1962.

19 "John C. Lindsay Dead, Solar Physics Expert," *Washington Post*, October 2, 1965, B3.

20 *Science Magazine*, March 16, 1962, 908.

21 NASA, "News media conference, OSO I spacecraft performance."

22 H.R. Brockett, Memorandum; "Urgent and temporary requirement for a command station in Libya to correct spurious emissions affecting OSO," April 10, 1962 (NASA History Division, Orbiting Solar Observatory files).

23 NASA, SP-57, 186.

24 "OSO Again Sending Data to the Sun," NASA News Release 62-152, July 1, 1962; launch spin rate of 120 RPM from NASA SP-57, 12.

25 "Radiochemistry Society": U.S. Nuclear Tests (Starfish Prime) http://www.radiochemistry.org/history/nuke_tests/dominic/index.html, (accessed July 4, 2007); "Degraded operations": John Simpson memo to Ball Brothers Research personnel, January 7, 1963; "Final shutoff": National Space Science Data Center, http://nssdc.gsfc.nasa.gov/database/MasterCatalog?sc=1962-006A (accessed July 4, 2007).

26 "NASA Facts: Orbiting Solar Observatory, Vol. 3, No. 7" (Washington, D.C.: U.S. Government Printing Office, 1966); Werner Neupert telephone interview, July 1, 2008; Ernest J. Ott, "Memorandum for chief of flight systems: Comments on the importance of OSO I," December 18, 1962 (NASA History Division, Orbiting Solar Observatory files).

27 "Ball of Fire," *Business Week.*

28 "NASA OSO Mismanagement Charged," *Space Daily,* February 7, 1964, 208; "Space Agency Hit by GAO, *Associated Press,* February 6, 1964.

29 Bester, *Life and Death of a Satellite,* 211.

Chapter 8

1 John Simpson, transcript of interview with David DeVorkin, July 28, 1983, Smithsonian National Air and Space Museum, *Space Astronomy Oral History Project*, 41.

[2] George Low, letter to Gordon Allott, December 22, 1970 (NASA History Division, Orbiting Solar Observatory files).

[3] W. David Compton and Charles D. Benson, *Living and Working in Space: A History of Skylab* (Washington, D.C.: NASA Scientific and Technical Information Branch, 1983), 70.

[4] "Cool Cosmos: IRAS Discoveries," http://coolcosmos.ipac.caltech.edu/cosmic_classroom/cosmic_reference/iras_discoveries.html (accessed January 10, 2007).

[5] "NASA/IPAC Infrared Science Archive: Introduction to IRAS," http://irsa.ipac.caltech.edu/IRASdocs/iras.html (accessed January 11, 2008).

[6] Eric J. Chaisson, *The Hubble Wars: Astrophysics Meets Astropolitics in the Two-billion-dollar Struggle Over the Hubble Space Telescope* (New York: HarperCollins, 1994), 29.

[7] John Noble Wilford, "Shuttle Soars 381 Miles High, With Telescope and a Dream," *New York Times*, April 25, 1990, 1A.

[8] Cassegrain reflector telescopes such as Hubble work by capturing light in a slightly concave donut of a primary mirror and bouncing it up to a slightly convex (think the back of a spoon) secondary mirror. The secondary mirror then bounces the light back through the hole in the center of the primary mirror to instruments waiting behind.

[9] David E. Chandler, "Manufacturing error leaves space telescope out of focus," *Boston Globe*, June 28, 1990, 1A.

[10] Chaisson, *Hubble Wars*, 185.

[11] Joseph Tatarewicz, "The Hubble Space Telescope Servicing Mission," in Pamela E. Mack, Ed., *From Engineering Science to Big Science: The NACA and NASA Collier Trophy Research Project Winners*, NASA History SP-4219 (Washington, D.C.: NASA History Office, 1998), 373.

[12] Chaisson, *Hubble Wars*, 184.

[13] Corrective Optics Space Telescope Axial Replacement (COSTAR) information packet, Ball Aerospace and Communications Group, June 1993, 10; Robert Brown and Holland Ford, Eds., "Report of the HST Strategy Panel: A Strategy for Recovery," Available at ftp://ftp.stsci.edu/pub/ExInEd/electronic_reports_folder/recovery.pdf, 35-72.

[14] Tatarewicz, "The Hubble Space Telescope Servicing Mission," 376.

[15] Robert Brown and Holland Ford, "Report of the HST Strategy Panel"; James H. Crocker, e-mail to author, March 22, 2010.

[16] COSTAR information packet, Ball Aerospace, 18-19.

[17] John Noble Wilford, "Space Telescope Confirms Theory of Black Holes," *New York Times*, May 26, 1994, 1A.

[18] Ron Young, personal interview, January 9, 2008.

[19] Among other examples of Dick Woolley's works include a filter for a space refrigerator capable of blocking gases while allowing liquids to pass (his inspiration being a bathtub washcloth holding a bubble of air); a "magnetotropometer" (a term of his invention) to measure a spacecraft's susceptibility to magnetic jostling, which involved a vat of silicon oil the size of a hot tub, a floating fiberglass donut and a two-story building held together with aluminum nails; a "great blue ox" machine to pull on space-shuttle clamps with 60,000 pounds of force; and various other technologies involving "spectacular pieces of apparatus," as Woolley put it.

[20] Lynn Schwertfeger, "High-flying Executive," *Boulder Daily Camera*, January 28, 1973.

[21] Until 2007, when the Wings Over the Rockies Air & Space Museum in Denver brought the Jetwing back to Denver, Colorado. Bartoe and wife Mary, with help from daughters Sally Palmer and Carolyn Pitts, reassembled the craft that July. It became a centerpiece of the museum.

[22] Michael Brett, "He's Sailing Away on a Jet Wing," *Auckland Star* (New Zealand), April 11, 1979.

[23] Ball advertisement clippings binder, Minnetrista Heritage Collection.

[24] Susan Smith, "Ball to buy Broomfield building," *Boulder Daily Camera*, August 5, 1987, 6D.

[25] Todd Neff, "50 years of exploration for Ball," *Boulder Daily Camera*, November 19, 2006, 1A.

Chapter 9

[1] Peter J. Westwick, *Into the Black: JPL and the American Space Program*, 1976–2004 (New Haven: Yale University Press, 2007), 264.

[2] William J. Broad, "NASA Moves to End Longtime Reliance on Big Spacecraft," *New York Times*, September 16, 1991, 1A.

[3] "National Aeronautics and Space Administration Small Planetary Mission Plan, Report to Congress," April 1992 (NASA History Division, Discovery folder), 1.

[4] Ibid., 2.

[5] Ibid., 3.

[6] Howard McCurdy, *Faster, Better, Cheaper: Low-cost Innovation in the U.S. Space Program* (Baltimore: The Johns Hopkins University Press, 2001), 50.

[7] Ibid., 51.

[8] "NASA Discovery Program Workshop Final Report," November 16–20, 1992, San Juan Capistrano Research Institute, San Juan Capistrano, California, 33.

[9] "NASA Selects 11 Discovery Mission Concepts for Study," NASA press release 93-027, February 11, 1993.

[10] Michael S. Belton, "New Frontiers in the Solar System: An Integrated Exploration Strategy Solar System Exploration Survey" (Washington, D.C.: National Academies Press, 2003), vii.

[11] Nigel Calder, *Giotto to the Comets* (London: Presswork, 1992), 112.

[12] Rich Reinert, personal interview, September 27, 2007.

[13] Alan Delamere, "Memorandum: Discovery Workshop in San Juan Capistrano, 16–18 November 1992," December 3, 1992 (Alan Delamere files).

[14] See HJ Melosh, *Impact Cratering: A Geologic Process* (New York: Oxford University Press, 1989).

Chapter 10

[1] Comet theorist Fred Whipple, who came up with the idea that comets had the constitutions of dirty snowballs, made the connection. The Geminids are named for Gemini, the constellation from which the shower's streaking meteoroids seem to emanate each December.

[2] Michael Belton, "Deep Impact: An Exploration of the Surface and Deep Sub-surface Structure of a Cometary Nucleus," NASA Discovery Program proposal, December 5, 1996, 2.

Chapter 11

[1] Calder, *Giotto to the Comets*, 61-62.

[2] Don Yeomans, telephone interview, February 26, 2008.

[3] "Comet Hale-Bopp: The Great Comet of 1997," Jet Propulsion Laboratory, http://www2.jpl.nasa.gov/comet/ (accessed January 19, 2008).

[4] Douglas Isbell, "Five Discovery Mission Proposals Selected for Feasibility Studies," NASA press release 98-203, November 12, 1998.

Chapter 12

[1] Hardware and associated numbers here and below are from Deep Impact project presentations. From the Preliminary Design Review of February 26–March 1, 2001: (1) Brian Muirhead, "Project Implementation Changes from CSR"; (2) Mike A'Hearn, "Deep Impact: The Science"; (3) William E. Frazier, "Technical Changes Since CSR." From the Critical Design Review of January 29–31, 2002: William E. Frazier, Kenny Epstein and Jeremy Stober, "6.0—Flight System Overview."

Chapter 13

1 Brian K. Muirhead and William L. Simon, *High Velocity Leadership* (New York: HarperCollins Business, 1999), 84.

2 Clayton R. Koppes, *JPL and the American Space Program: A History of the Jet Propulsion Laboratory* (New Haven: Yale University Press, 1982), 2-20.

3 Westwick, *Into the Black*, IX.

4 Ibid., 308.

5 McCurdy, *Faster, Better, Cheaper*, 81.

6 Westwick, *Into the Black*, 278.

7 Paul Hoversten, "Bad Math Added Up to Doomed Mars Spacecraft," *USA Today*, October 1, 1999.

8 Dan Rather, "NASA Needs a Wake-up Call—and This Is It," *Houston Chronicle*, April 9, 2000, 6.

9 Matthew Fordahl, "NASA Chief Takes Blame," *Associated Press*, March 30, 2000.

Chapter 15

1 Mark Rayman, "Dr. Mark Rayman's Mission Log," September 23, 2008, http://nmp.jpl.nasa.gov/ds1/arch/mrlog72.html (accessed March 7, 2008).

2 Donald L. Hampton et al, "An Overview of the Instrument Suite for the Deep Impact Mission," *Space Science Reviews* (2005) 117: 43-93, 57.

3 L.A. Soderblom, et al, "Short-wavelength infrared (1.3–2.6 µm) observations of the nucleus of Comet 19P/Borrelly," *Icarus* 167 (2004) 100–112, 108.

4 Carbon, cyanide ions, and hydroxyls in particular. See Donald L. Hampton et al, "An Overview of the Instrument Suite for the Deep Impact Mission," 55.

Chapter 17

1 NASA, "Comet Nucleus Tour Mishap Investigation Board Report," May 31, 2003, 12-14.

2 NASA, "Columbia Accident Investigation Board Report, Volume 1," August 2003, 191.

Chapter 18

1 Carolyn S. Griner and W. Brian Keegan, "Enhancing Mission Success—A Framework for the Future," a report by the NASA chief engineer and the NASA integrated action team, December 21, 2000, 46.

Chapter 20

1 Chris Burno, personal interview, October 15, 2007.

2 Tom Golden, personal interview, November 5, 2007.

Chapter 21

1 Tom Bank, personal interview, September 27, 2005.

2 Chris Burno, personal interview; Brian Buchholtz, Deep Impact flight system pre-ship review, "Structures and Mechanisms," 8.

3 Shari Apslund, "Re. Native American culture," e-mail communication, Friday, June 18, 2004.

4 Monte Henderson, "Re. Native American culture," e-mail communication, Sunday, June 20, 2004.

5 Rick Grammier, "Re. Native American culture," e-mail communication, Thursday, June 17, 2004.

6 Mike A'Hearn, letter to Lucy McFadden, July 2, 2004.

7 Katy Human, "Boulder-built comet collider aims for science home run," *Denver Post*, October 3, 2004, C1.

Chapter 22

[1] K.J. Barltrop, and E.P. Kan, "How much fault protection is enough: A Deep Impact perspective," Aerospace Conference, 2005 IEEE, March 5-12 2005: 1-14, 11.

[2] "Interview with NASA Launch Manager Omar Baez," Stream Video KSC-05-S-00026, January 14, 2005.

Chapter 23

[1] Glen Roat, "Deep Impact Post Launch Assessment Review Assessment Report," 18.

[2] Kevin Barltrop, personal interview, May 4, 2009.

[3] Wikipedia, "List of Care Bears," http://en.wikipedia.org/wiki/List_of_Care_Bears (Accessed April 18, 2008).

[4] James Fanson, Jim Oschmann, and Michael A'Hearn, "HRI Focus Anomaly Tiger Team Investigation: Final Report," December 15, 2005, p. 7.

[5] "Deep Impact Mission Status Report," NASA press release 05-086, March 25, 2005.

[6] Jim Erickson, "Boulder spacecraft closes in on comet," *Rocky Mountain News*, April 28, 2005, 24A.

[7] Chaisson, *Hubble Wars*, 177.

[8] Ibid., 179.

Chapter 24

[1] Rudyard Kipling, "If," Fordham University Modern History Sourcebook, available at http://www.fordham.edu/halsall/mod/kipling-if.html.

[2] Daniel G. Kubitschek et al, "The Challenges of Deep Impact Autonomous Navigation," *Journal of Field Robotics* 24(4), 339-354 (2007).

[3] Jessica M. Sunshine et al., "Expectations for infrared spectroscopy of 9P/Tempel 1 from Deep Impact," *Space Science Reviews* (2005) 117: 269-295. For chemicals spotted in comets, see table on p. 283.

[4] "NASA Announced Spectacular Day of the Comet," NASA press release 05-144, June 9, 2005.

Epilogue

[1] "Collision with a comet," *New York Times*, July 6, 2005. Available online at http://www.nytimes.com/2005/07/06/opinion/06wed3.html (accessed September 8, 2009).

[2] "Spectacular Impact Draws Millions to NASA Portal," NASA press release 05-174, July 6, 2005.

[3] J.E Richardson et al., "A ballistics analysis of the Deep Impact ejecta plume: Determining Comet Tempel 1's gravity, mass, and density," *Icarus* 190 (2007) 357-390.

[4] "NASA Successfully Tests First Deep Space Internet," NASA press release 08-298, November 18, 2008.

[5] "Space Scope Over State Predicted," *Prescott Evening Courier*, July 17, 1959.

[6] Rick Shaffer, telephone interview, July 20, 2007.

[7] "Tribute to Edmund F. Ball," Congressional Record—Senate (S10247), October 11, 2000.

Note Regarding Sources
With a few exceptions, I have omitted endnote citations relating to direct quotes from personal and telephone interviews as well as to information from proprietary Ball Aerospace documents. Interview sources are usually clear from context (that is, if it's a quote and it's not footnoted, I talked to them). The Ball Aerospace documents provided technical detail related to the Deep Impact spacecraft, and included hundreds of Microsoft PowerPoint presentations and other files from Deep Impact's Preliminary Design Review, Critical Design Review, System Test Readiness Review, Environmental Test Readiness Review, CEO Reviews, Pre-Ship Review, Risk Reviews, Mission Readiness Review, Post-Launch Assessment Review, Encounter Risk Review, Critical Events Readiness Review, weekly status reports from 2003–2005, and Discovery Program quarterly reviews.

ACKNOWLEDGMENTS

MANY, MANY PEOPLE helped make this book happen. My wife Carol gave her patience and support. Kevin Kaufman, the Boulder *Daily Camera*'s editor, tolerated my late-night rummaging through the newspaper's archives and backed my fellowship at the University of Colorado, during which I was able to focus on this effort.

CU journalism professors Len Ackland and Tom Yulsman looked the other way as I worked on a space engineering book during a good portion of my Ted Scripps Fellowship in Environmental Journalism, without which I could have never told this story.

Monte Henderson opened the doors to Deep Impact. Ball Aerospace's media relations team, including Sarah Sloan, Roz Brown, and Emilia Reed, kept them ajar. Scott McCarty, head of media relations for Ball Corporation, was equally supportive, both in terms of archival access and the use of images. JPL Public Information Officer D.C. Agle was also a big help in securing time with JPL sources.

Bob Arentz cost me several months of my life. I had been mainly interested in how a modern space mission plays out and would not have considered pursuing the Ball Aerospace's history—nor that of University of Colorado Laboratory for Atmospheric and Space Physics—had it not been for our conversations.

Stephanie and Jeanne Safford sent me helpful documents from the Control Cells era. Emily CoBabe-Ammann happened to have a set of University of Colorado Upper Air Laboratory reports to the Air Force on her LASP office shelves, left in her care by a recent retiree. On multiple occasions, she availed her office to my scanner, laptop and digital camera. Alan Delamere shed much light on the early days of the Discovery Program, the Deep Impact proposal process and the business of building spacecraft in general.

I'm indebted to Susan Smith, archivist at the Minnetrista Heritage Collection in Muncie, Indiana, whose generous research and assistance was indispensable in untangling the origins of Ball Brothers Research.

I have fond memories of a December weekend spent with Pete and Mary Bartoe in their home in Clark, Colorado, and of sitting in the cockpit the Jetwing at the Wings Over the Rockies Air & Space Museum in Denver.

Mike Cote, Marissa and Barry Elk, Matt Kelly, Vickie Michaelis, Steve Still, Chris Young, and my brother Chris Neff provided valuable input on various drafts. Rob Ford and Greg Laugero offered vital feedback on final drafts. My mom Kay Neff was a skilled and tireless copy editor. My old friend David Barringer edited and designed this book, and then shepherded it through the publishing process.

I'm grateful to Ben Bova, who offered insightful comments on early drafts, and to his wife Barbara Bova. Barbara, who succumbed to cancer in 2009, did all she could to see that this book came to be.

I garnered valuable insights from hundreds of photographs Ben Toyoshima took at the Jet Propulsion Laboratory during the Deep Impact effort. Mike Hughes, an indispensable source on the comet mission from the JPL perspective, provided discs of Ben's photos as well as many other images and documents that filled important blanks.

Chase Hutto kindly hosted me during a Washington D.C. reporting trip. My mother-in-law, Joe Anna Eirich, was a faithful transcriber of many, many interviews—not a few of them done in Boulder's Conor O'Neill's Pub, for whose Guinness I remain thankful.

Finally, more than 100 people provided insights and helped bring the Ball Aerospace and Deep Impact stories to life in personal, phone and e-mail interviews: Mike A'Hearn, David Acton, Len Andreozzi, Bob Arentz, Jim Badger, Jim Baer, Alec Baldwin, Tom Bank, Kevin Barltrop, Albert Bartlett, O.E. "Pete" Bartoe, Mary Bartoe, Joseph Bassi, William Behring, Mike Belton, Bill Blume, Sandy Bracken, Donald Brownlee, Carl Buck, Chris Burno, Palmer Carlin, Nancy Carlson, Jerry Chodil, Phil Christensen, Earl Conn, Chuck Cornelison, James Crocker, Alan Delamere, Barbara Dolder, Karl Dolder, Dennis Ebbetts, Sean Esslinger, John Fisher, Bill Frank, Bill Frazier, Marion Fulk, Genevieve Gablehouse, Joe Galamback, Stamer Geminden, Tom Golden, Michele Goldman, Robert Goldstein, Rick Grammier, Don Hampton, Monte Henderson, Lorna Hess-Frey, John Houlton, Mike Hughes, Marty Huisjen, Ken Hutchinson, Vivian Jackson, Lindley Johnson, Lew Kendall, Marcy Kendall, John King, Dan Kubitschek, Carol Lane, Steve Ling, Jon Mah, John Marriott, Nick Mastrodemos, Michael McLelland, John McNamee, Karen Meech, Jay Melosh, Ruel C. "Merc" Mercure, Harold Montoya, Marilyn Morris, Brian Muirhead, Larry Murphy, Marcia Neugebauer, Werner Neupert, James Nidey, Tim Ostwald, Laurence Paddock, Keyur Patel, Alice Phinney, William J. Pietenpol, Sean Raymond, Emilia Reed, Rich Reinert, Merle Reisbeck, Harold Reitsema, William Rense, Jennifer Rocca, Nancy Grace Roman, Tomas Ryan, Stephanie Safford, Charlie Schira, Peter Schultz, Rick Shaffer, William Simons, Jim Snow, Steve Sodja, Tom Spencer, Kim Stacey, Peter Stacey, Wayne Stacey, Stephanie Stafford, Alan Stern, Jeremy Stober, Jessica Sunshine, David Taylor, Nick Taylor, Scott Tibbitts, John Trauger, John Valdez, Joe Veverka, Buddy Walls, Terry Wallace, Amy Walsh, Ed Weiler, Peter Westwick, Joel Williamsen, David Wilson, Dick Woolley, Tom Yarnell, Don Yeomans, and Ron Young.

Many generously spent hours talking with me, often on more than one occasion. I can't thank them enough.

PHOTO CREDITS

Ball Aerospace & Technologies Corporation: page 51, 54, 64, 65, 94, 110, 112, 130, 136, 140, 148, 149, 172, 192, 198, 200, 202, 207, 213, 217, 224, 231, 234, 237, 240, 241, 253, 267 (top), 278, 313, 315; Ball Corporation: 37, 38, 39, 41, 44; European Space Agency, courtesy of MPAe, Lindau: 119; *The Daily Camera*: 15, 36; Mike Keefe: 309; NASA: 75, 88, 90, 106, 141, 267 (bottom); NASA/JPL-Caltech: 184, 271, 295, 298, 303, 305, 307; Werner Neupert: 59; Ben Toyoshima: 132, 188, 218, 262, 272, 273, 275; The University of Colorado Laboratory for Atmospheric and Space Physics: 18, 22, 23, 25, 26, 28, 30, 31; Joel Williamsen: 182; Don Yeomans: 154.

INDEX